Studer
Solutions Manual

An Introduction to
Mathematical
Statistics
and Its Applications
Fourth Edition

Richard J. Larsen | Morris L. Marx
Vanderbilt University | University of West Florida

PEARSON
Prentice
Hall

Upper Saddle River, NJ 07458

Executive Acquisitions Editor: Petra Recter
Editor-in-Chief: Sally Yagan
Supplement Editor: Joanne Wendelken
Executive Managing Editor: Kathleen Schiaparelli
Assistant Managing Editor: Karen Bosch
Production Editor: Robert Merenoff
Supplement Cover Manager: Paul Gourhan
Supplement Cover Designer: Christopher Kossa
Manufacturing Buyer: Ilene Kahn
Manufacturing Manager: Alexis Heydt-Long

© 2006 Pearson Education, Inc.
Pearson Prentice Hall
Pearson Education, Inc.
Upper Saddle River, NJ 07458

Printed in the United States of America

10 9 8 7 6 5 4 3 2 1

ISBN 0-13-186796-2

Pearson Education Ltd., London
Pearson Education Australia Pty. Ltd., Sydney
Pearson Education Singapore, Pte. Ltd.
Pearson Education North Asia Ltd., Hong Kong
Pearson Education Canada, Inc., Toronto
Pearson Educación de Mexico, S.A. de C.V.
Pearson Education—Japan, Tokyo
Pearson Education Malaysia, Pte. Ltd.

Table of Contents

Chapter 2

Section 2.2

2.2.1 $S = \{(s,s,s), (s,s,f), (s,f,s), (f,s,s), (s,f,f), (f,s,f), (f,f,s), (f,f,f)\}$
$A = \{(s,f,s), (f,s,s)\}; \quad B = \{(f,f,f)\}$

2.2.3 $(1,3,4), (1,3,5), (1,3,6), (2,3,4), (2,3,5), (2,3,6)$

2.2.5 The outcome sought is $(4, 4)$. It is "harder" to obtain than the set $\{(5, 3), (3, 5), (6, 2), (2, 6)\}$ of other outcomes making a total of 8.

2.2.7 $P = \{\text{right triangles with sides } (5, a, b): a^2 + b^2 = 25\}$

2.2.9 (a) $S = \{(0, 0, 0, 0) (0, 0, 0, 1), (0, 0, 1, 0), (0, 0, 1, 1), (0, 1, 0, 0), (0, 1, 0, 1), (0, 1, 1, 0),$
$(0, 1, 1, 1), (1, 0, 0, 0), (1, 0, 0, 1), (1, 0, 1, 0), (1, 0, 1, 1,), (1, 1, 0, 0), (1, 1, 0, 1),$
$(1, 1, 1, 0), (1, 1, 1, 1,)\}$

       (b) $A = \{(0, 0, 1, 1), (0, 1, 0, 1), (0, 1, 1, 0), (1, 0, 0, 1), (1, 0, 1, 0), (1, 1, 0, 0,)\}$

       (c) $1 + k$

2.2.11 Let p_1 and p_2 denote the two perpetrators and i_1, i_2, and i_3, the three in the lineup who are innocent. Then
$S = \{(p_1,i_1), (p_1,i_2), (p_1,i_3), (p_2,i_1), (p_2,i_2), (p_2,i_3), (p_1,p_2), (i_1,i_2), (i_1,i_3), (i_2,i_3)\}$
The event A contains every outcome in S except (p_1, p_2).

2.2.13 In order for the shooter to win with a point of 9, one of the following (countably infinite) sequences of sums must be rolled: (9,9), (9, no 7 or no 9,9), (9, no 7 or no 9, no 7 or no 9,9), …

2.2.15 Let A_k be the set of chips put in the urn at $1/2^k$ minute until midnight. For example, $A_1 = \{11, 12, 13, 14, 15, 16, 17, 18, 19, 20\}$.

Then the set of chips in the urn at midnight is $\bigcup_{k=1}^{\infty}(A_k - \{k+1\}) = \varnothing$

2.2.17 If $x^2 + 2x \leq 8$, then $(x + 4)(x - 2) \leq 0$ and $A = \{x: -4 \leq x \leq 2\}$. Similarly, if $x^2 + x \leq 6$, then $(x + 3)(x - 2) \leq 0$ and $B = \{x: -3 \leq x \leq 2\}$. Therefore, $A \cap B = \{x: -3 \leq x \leq 2\}$ and $A \cup B = \{x: -4 \leq x \leq 2\}$.

2.2.19 The system fails if either the first pair fails or the second pair fails (or both pairs fail). For either pair to fail, though, both of its components must fail. Therefore,
$A = (A_{11} \cap A_{21}) \cup (A_{12} \cap A_{22})$.

2.2.21 40

2.2.23 (a) If s is a member of $A \cup (B \cap C)$ then s belongs to A or to $B \cap C$. If it is a member of A or of $B \cap C$, then it belongs to $A \cup B$ and to $A \cup C$. Thus, it is a member of $(A \cup B) \cap (A \cup C)$. Conversely, choose s in $(A \cup B) \cap (A \cup C)$. If it belongs to A, then it belongs to $A \cup (B \cap C)$. If it does not belong to A, then it must be a member of $B \cap C$. In that case it also is a member of $A \cup (B \cap C)$.

(b) If s is a member of $A \cap (B \cup C)$ then s belongs to A and to $B \cup C$. If it is a member of B, then it belongs to $A \cap B$ and, hence, $(A \cap B) \cup (A \cap C)$. Similarly, if it belongs to C, it is a member of $(A \cap B) \cup (A \cap C)$. Conversely, choose s in $(A \cap B) \cup (A \cap C)$. Then it belongs to A. If it is a member of $A \cap B$ then it belongs to $A \cap (B \cup C)$. Similarly, if it belongs to $A \cap C$, then it must be a member of $A \cap (B \cup C)$.

2.2.25 (a) Let s be a member of $A \cup (B \cup C)$. Then s belongs to either A or $B \cup C$ (or both). If s belongs to A, it necessarily belongs to $(A \cup B) \cup C$. If s belongs to $B \cup C$, it belongs to B or C or both, so it must belong to $(A \cup B) \cup C$. Now, suppose s belongs to $(A \cup B) \cup C$. Then it belongs to either $A \cup B$ or C or both. If it belongs to C, it must belong to $A \cup (B \cup C)$. If it belongs to $A \cup B$, it must belong to either A or B or both, so it must belong to $A \cup (B \cup C)$.

(b) Suppose s belongs to $A \cap (B \cap C)$, so it is a member of A and also $B \cap C$. Then it is a member of A and of B and C. That makes it a member of $(A \cap B) \cap C$. Conversely, if s is a member of $(A \cap B) \cap C$, a similar argument shows it belongs to $A \cap (B \cap C)$.

2.2.27 A is a subset of B.

2.2.29 (a) B and C
(b) B is a subset of A.

2.2.31 Let A and B denote the students who saw the movie the first time and the second time, respectively. Then $N(A) = 850$, $N(B) = 690$, and $N((A \cup B)^C) = 4700$ (implying that $N(A \cup B) = 1300$). Therefore, $N(A \cap B) =$ number who saw movie twice $= 850 + 690 - 1300 = 240$.

2.2.33 (a)

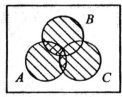

$A \cap (B \cup C) = (A \cap B) \cup (A \cap C)$

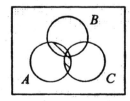

(b)

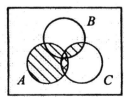

$A \cup (B \cap C) = (A \cup B) \cap (A \cup C)$

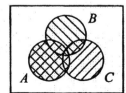

2.2.35 A and B are subsets of $A \cup B$.

2.2.37 Let A be the set of those with MCAT scores ≥ 27 and B be the set of those with GPAs ≥ 3.5. We are given that $N(A) = 1000$, $N(B) = 400$, and $N(A \cap B) = 300$. Then
$$N(A^C \cap B^C) = N[(A \cup B)^C] = 1200 - N(A \cup B)$$
$$= 1200 - [(N(A) + N(B) - N(A \cap B)]$$
$$= 1200 - [(1000 + 400 - 300] = 100.$$
The requested proportion is 100/1200.

2.2.39 Let A be the set of those saying "yes" to the first question and B be the set of those saying "yes" to the second question. We are given that $N(A) = 600$, $N(B) = 400$, and $N(A^C \cap B) = 300$. Then $N(A \cap B) = N(B) - N(A^C \cap B) = 400 - 300 = 100$.
$N(A \cap B^C) = N(A) - N(A \cap B) = 600 - 100 = 500.$

Section 2.3

2.3.1 Let L and V denote the sets of programs with offensive language and too much violence, respectively. Then $P(L) = 0.42$, $P(V) = 0.27$, and $P(L \cap V) = 0.10$. Therefore, P(program complies) $= P((L \cup V)^C) = 1 - [P(L) + P(V) - P(L \cap V)] = 0.41$.

2.3.3 (a) $1 - P(A \cap B)$
(b) $P(B) - P(A \cap B)$

2.3.5 No. $P(A_1 \cup A_2 \cup A_3) = P$(at least one "6" appears) $= 1 - P$(no 6's appear) $= 1 - \left(\dfrac{5}{6}\right)^3 \neq \dfrac{1}{2}$.

The A_i's are not mutually exclusive, so $P(A_1 \cup A_2 \cup A_3) \neq P(A_1) + P(A_2) + P(A_3)$.

2.3.7

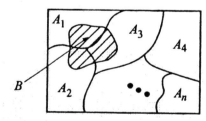

By inspection, $B = (B \cap A_1) \cup (B \cap A_2) \cup \ldots \cup (B \cap A_n)$.

2.3.9 P(odd man out) $= 1 - P$(no odd man out) $= 1 - P(HHH \text{ or } TTT) = 1 - \dfrac{2}{8} = \dfrac{3}{4}$.

2.3.11 Let A: State wins Saturday and B: State wins next Saturday. Then $P(A) = 0.10$, $P(B) = 0.30$, and P(lose both) $= 0.65 = 1 - P(A \cup B)$, which implies that $P(A \cup B) = 0.35$. Therefore, $P(A \cap B) = 0.10 + 0.30 - 0.35 = 0.05$, so P(State wins exactly once) $= P(A \cup B) - P(A \cap B) = 0.35 - 0.05 = 0.30$.

2.3.13 Let F: female is hired and T: minority is hired. Then $P(F) = 0.60$, $P(T) = 0.30$, and $P(F^C \cap T^C) = 0.25 = 1 - P(F \cup T)$. Since $P(F \cup T) = 0.75$, $P(F \cap T) = 0.60 + 0.30 - 0.75 = 0.15$.

2.3.15 (a) $X^C \cap Y = \{(H, T, T, H), (T, H, H, T)\}$, so $P(X^C \cap Y) = 2/16$

(b) $X \cap Y^C = \{(H, T, T, T), (T, T, T, H), (T, H, H, H), (H, H, H, T)\}$ so $P(X \cap Y^C) = 4/16$

2.3.17 $A \cap B, (A \cap B) \cup (A \cap C), A, A \cup B, S$

Section 2.4

2.4.1 $P(\text{sum} = 10 \mid \text{sum exceeds } 8) = \dfrac{P(\text{sum} = 10 \text{ and sum exceeds } 8)}{P(\text{sum exceeds } 8)} =$

$\dfrac{P(\text{sum} = 10)}{P(\text{sum} = 9, 10, 11, \text{ or } 12)} = \dfrac{3/36}{4/36 + 3/36 + 2/36 + 1/36} = \dfrac{3}{10}$.

2.4.3 If $P(A \mid B) = \dfrac{P(A \cap B)}{P(B)} < P(A)$, then $P(A \cap B) < P(A) \cdot P(B)$. It follows that

$P(B \mid A) = \dfrac{P(A \cap B)}{P(A)} < \dfrac{P(A) \cdot P(B)}{P(A)} = P(B)$.

2.4.5 The answer would remain the same. Distinguishing only three family types does not make them equally likely; (girl, boy) families will occur twice as often as either (boy, boy) or (girl, girl) families.

2.4.7 Let R_i be the event that a red chip is selected on the ith draw, $i = 1, 2$. Then $P(\text{both are red}) =$

$P(R_1 \cap R_2) = P(R_2 \mid R_1)P(R_1) = \dfrac{3}{4} \cdot \dfrac{1}{2} = \dfrac{3}{8}$.

2.4.9 Let W_i be the event that a white chip is selected on the ith draw, $i = 1,2$. Then $P(W_2 \mid W_1) = \dfrac{P(W_1 \cap W_2)}{P(W_1)}$. If both chips in the urn are white, $P(W_1) = 1$; if one is white and one is black,

$P(W_1) = \dfrac{1}{2}$. Since each chip distribution is equally likely, $P(W_1) = 1 \cdot \dfrac{1}{2} + \dfrac{1}{2} \cdot \dfrac{1}{2} = \dfrac{3}{4}$.

Similarly, $P(W_1 \cap W_2) = 1 \cdot \dfrac{1}{2} + \dfrac{1}{4} \cdot \dfrac{1}{2} = \dfrac{5}{8}$, so $P(W_2 \mid W_1) = \dfrac{5/8}{3/4} = \dfrac{5}{6}$.

2.4.11 (a) $P(A^C \cap B^C) = 1 - P(A \cup B)$
$= 1 - [P(A) + P(B) - P(A \cap B)]$
$= 1 - [0.65 + 0.55 - 0.25] = 0.05$

(b) $P[(A^C \cap B) \cup (A \cap B^C)] = P(A^C \cap B) + P(A \cap B^C)$
$= [P(A) - P(A \cap B)] + [P(B) - P(A \cap B)]$
$= [0.65 - 0.25] + [0.55 - 0.25] = 0.70$

(c) $P(A \cup B) = 0.95$

(d) $P[(A \cap B)^C] = 1 - P(A \cap B) = 1 - 0.25 = 0.75$

(e) $P\{[(A^C \cap B) \cup (A \cap B^C)] \mid A \cup B\}$

$$= \frac{P[(A^C \cap B) \cup (A \cap B^C)]}{P(A \cup B)} = 0.70/0.95 = 70/95$$

(f) $P(A \cap B) \mid A \cup B) = P(A \cap B)/P(A \cup B) = 0.25/0.95 = 25/95$

(g) $P(B \mid A^C) = P(A^C \cap B)/P(A^C)] = [P(B) - P(A \cap B)]/[1 - P(A)]$
$$= [0.55 - 0.25]/[1 - 0.65] = 30/35$$

2.4.13 $P(\text{first die} \geq 4 \mid \text{sum} = 8)$
$$= P(\text{first die} \geq 4 \text{ and sum} = 8)/P(\text{sum} = 8)$$
$$= P(\{(4, 4), (5, 3), (6, 2)\})/P(\{(2, 6), (3, 5), (4, 4), (5, 3), (6, 2)\}) = 3/5$$

2.4.15 First note that $P(A \cup B) = 1 - P[(A \cup B)^C] = 1 - 0.2 = 0.8$.
Then $P(B) = P(A \cup B) - P(A \cap B^C) - P(A \cap B) = 0.8 - 0.3 - 0.1 = 0.5$. Finally $P(A \mid B) = P(A \cap B)/P(B) = 0.1/0.5 = 1/5$

2.4.17 $P[(A \cap B)^C] = P[(A \cup B)^C] + P(A \cap B^C) + P(A^C \cap B) = 0.2 + 0.1 + 0.3 = 0.6$
$P(A \cup B \mid (A \cap B)^C) = P[(A \cap B^C) \cup (A^C \cap B)]/P((A \cap B)^C) = [0.1 + 0.3]/0.6 = 2/3$

2.4.19 $P(\text{Outandout wins} \mid \text{Australian Doll and Dusty Stake don't win}) = P(\text{Outandout wins and Australian Doll and Dusty Stake don't win})/P(\text{Australian Doll and Dusty Stake don't win}) = 0.20/0.55 = 20/55$

2.4.21 $P(BBRWW) = P(B)P(B \mid B)P(R \mid BB)P(W \mid BBR)P(W \mid BBRW) = \dfrac{4}{15} \cdot \dfrac{3}{14} \cdot \dfrac{5}{13} \cdot \dfrac{6}{12} \cdot \dfrac{5}{11} =$

0.0050. $P(2, 6, 4, 9, 13) = \dfrac{1}{15} \cdot \dfrac{1}{14} \cdot \dfrac{1}{13} \cdot \dfrac{1}{12} \cdot \dfrac{1}{11} = \dfrac{1}{360,360}$.

2.4.23 $(1/52)(1/51)(1/50)(1/49) = 1/6,497,400$

2.4.25 Let A_i be the event "Bearing came from supplier i", $i = 1, 2, 3$. Let B be the event "Bearing in toy manufacturer's inventory is defective." Then
$$P(A_1) = 0.5, P(A_2) = 0.3, P(A_3 = 0.2)$$
and
$$P(B \mid A_1) = 0.02, P(B \mid A_2) = 0.03, P(B \mid A_3) = 0.04$$
Combining these probabilities according to Theorem 2.4.1 gives
$$P(B) = (0.02)(0.5) + (0.03)(0.3) + (0.04)(0.2)$$
$$= 0.027$$
meaning that the manufacturer can expect 2.7% of her ball-bearing stock to be defective.

2.4.27 Let B be the event that the countries go to war. Let A be the event that terrorism increases. Then $P(B) = P(B \mid A)P(A) + P(B \mid A^C)P(A^C) = (0.65)(0.30) + (0.05)(0.70) = 0.23$.

2.4.29 Let B denote the event that the person interviewed answers truthfully, and let A be the event that the person interviewed is a man. Then $P(B) = P(B \mid A)P(A) + P(B \mid A^C)P(A^C) = (0.78)(0.47) + (0.63)(0.53) = 0.70$.

Chapter 2

5

2.4.31 Let B denote the event that the attack is a success, and let A denote the event that the Klingons interfere. Then $P(B) = P(B|A)P(A) + P(B|A^C)P(A^C) = (0.3)(0.2384) + (0.8)(0.7616) = 0.68$. Since $P(B) < 0.7306$, they should not attack.

2.4.33 If B is the event that Backwater wins and A is the event that their first-string quarterback plays, then $P(B) = P(B|A)P(A) + P(B|A^C)P(A^C) = (0.75)(0.70) + (0.40)(0.30) = 0.645$.

2.4.35 No. Let B denote the event that the person calling the toss is correct. Let A_H be the event that the coin comes up Heads and let A_T be the event that the coin comes up Tails. Then $P(B) =$

$$P(B|A_H)P(A_H) + P(B|A_T)P(A_T) = (0.7)\left(\frac{1}{2}\right) + (0.3)\left(\frac{1}{2}\right) = \frac{1}{2}.$$

2.4.37 Let A_1 be the event of a 3.5-4.0 GPA; A_2, of a 3.0-3.5 GPA; and A_3, of a GPA less than 3.0. If B is the event of getting into medical school, then
$$P(B) = P(B|A_1)P(A_1) + P(B|A_2)P(A_2) + P(B|A_3)P(A_3)$$
$$= (0.8)(0.25) + (0.5)(0.35) + (0.1)(0.40) = 0.415$$

2.4.39 Let A_1 be the event of being a Humanities major; A_2, of being a Natural Science major; A_3, of being a History major; and A_4, of being a Social Science major. If B is the event of a male student, then
$$P(B) = P(B|A_1)P(A_1) + P(B|A_2)P(A_2) + P(B|A_3)P(A_3) + P(B|A_4)P(A_4)$$
$$= (0.40)(0.4) + (0.85)(0.1) + (0.55)(0.3) + (0.25)(0.2)$$
$$= 0.46$$

2.4.41 Let A_i be the event that Urn i is chosen, $i = $ I, II, III. Then, $P(A_i) = 1/3$, $i = $ I, II, III. Suppose B is the event a red chip is drawn. Note that $P(B|A_1) = 3/8$, $P(B|A_2) = 1/2$ and $P(B|A_3) = 5/8$.

$$P(A_3|B) = \frac{P(B|A_3)P(A_3)}{P(B|A_1)P(A_1) + P(B|A_2)P(A_2) + P(B|A_3)P(A_3)}$$
$$= \frac{(5/8)(1/3)}{(3/8)(1/3) + (1/2)(1/3) + (5/8)(1/3)} = 5/12.$$

2.4.43 Let B be the event that the basement leaks, and let A_T, A_W, and A_H denote the events that the house was built by Tara, Westview, and Hearthstone, respectively. Then $P(B|A_T) = 0.60$, $P(B|A_W) = 0.50$, and $P(B|A_H) = 0.40$. Also, $P(A_T) = 2/11$, $P(A_W) = 3/11$, and $P(A_H) = 6/11$. Applying Bayes' rule to each of the builders shows that $P(A_T|B) = 0.24$, $P(A_W|B) = 0.29$, and $P(A_H|B) = 0.47$, implying that Hearthstone is the most likely contractor.

2.4.45 Let B denote the event that a check bounces, and let A be the event that a customer wears sunglasses. Then $P(B|A) = 0.50$, $P(B|A^C) = 1 - 0.98 = 0.02$, and $P(A) = 0.10$, so

$$P(A|B) = \frac{(0.50)(0.10)}{(0.50)(0.10) + (0.02)(0.90)} = 0.74$$

Chapter 2

2.4.47 Define B to be the event that Josh answers a randomly selected question correctly, and let A_1 and A_2 denote the events that he was 1) unprepared for the question and 2) prepared for the question, respectively. Then $P(B|A_1) = 0.20$, $P(B|A_2) = 1$, $P(A_2) = p$, $P(A_1) = 1 - p$, and

$$P(A_2 | B) = 0.92 = \frac{P(B|A_2)P(A_2)}{P(B|A_1)P(A_1) + P(B|A_2)P(A_2)} = \frac{1 \cdot p}{(0.20)(1-p) + (1 \cdot p)}$$

which implies that $p = 0.70$ (meaning that Josh was prepared for $(0.70)(20) = 14$ of the questions).

2.4.49 Let A_1 be the event of being a Humanities major; A_2, of being a History and Culture major; and A_3, of being a Science major. If B is the event of being a woman, then

$$P(A_2 | B) = \frac{(0.45)(0.5)}{(0.75)(0.3) + (0.45)(0.5) + (0.30)(0.2)} = 225/510$$

2.4.51 Let B be the event that Zach's girlfriend responds promptly. Let A be the event that Zach sent an e-mail, so A^C is the event of leaving a message. Then

$$P(A | B) = \frac{(0.8)(2/3)}{(0.8)(2/3) + (0.9)(1/3)} = 16/25$$

2.4.53 Let A_i be the event that Drawer i is chosen, $i, = 1, 2, 3$. If B is the event a silver coin is selected, then

$$P(A_3 | B) = \frac{(0.5)(1/3)}{(0)(1/3) + (1)(1/3) + (0.5)(1/3)} = 1/3$$

Section 2.5

2.5.1 a) No, because $P(A \cap B) > 0$.
　　　　b) No, because $P(A \cap B) = 0.2 \neq P(A) \cdot P(B) = (0.6)(0.5) = 0.3$
　　　　c) $P(A^C \cup B^C) = P((A \cap B)^C) = 1 - P(A \cap B) = 1 - 0.2 = 0.8$.

2.5.3 $P(\text{one face is twice the other face}) = P((1, 2), (2, 1), (2, 4), (4, 2), (3, 6), (6, 3)) = \dfrac{6}{36}$.

2.5.5 $P(\text{Dana wins at least 1 game out of 2}) = 0.3$, which implies that $P(\text{Dana loses 2 games out of 2}) = 0.7$. Therefore, $P(\text{Dana wins at least 1 game out of 4}) = 1 - P(\text{Dana loses all 4 games}) = 1 - P(\text{Dana loses first 2 games} \underline{\text{ and}} \text{ Dana loses second 2 games}) = 1 - (0.7)(0.7) = 0.51$.

2.5.7 (a) 1. $P(A \cup B) = P(A) + P(B) - P(A \cap B) = 1/4 + 1/8 + 0 = 3/8$

2. $P(A \cup B) = P(A) + P(B) - P(A)P(B)$
$$= 1/4 + 1/8 - (1/4)(1/8) = 11/32$$

(b) 1. $P(A|B) = \dfrac{P(A \cap B)}{P(B)} = \dfrac{0}{P(B)} = 0$

2. $P(A|B) = \dfrac{P(A \cap B)}{P(B)} = \dfrac{P(A)P(B)}{P(B)} = P(A) = 1/4$

2.5.9 Let A_i be the event of i heads in the first two tosses, $i = 0, 1, 2$. Let B_i be the event of i heads in the last two tosses, $i = 0, 1, 2$. The A's and B's are independent. The event of interest is $(A_0 \cap B_0) \cup (A_1 \cap B_1) \cup (A_2 \cap B_2)$ and $P[(A_0 \cap B_0) \cup (A_1 \cap B_1) \cup (A_2 \cap B_2)] = P(A_0)P(B_0) + P(A_1)P(B_1) + P(A_2)P(B_2) = (1/4)(1/4) + (1/2)(1/2) + (1/4)(1/4) = 6/16$

2.5.11 Equation 2.5.3:
$$P(A \cap B \cap C) = P(\{1, 3)\}) = 1/36 = (2/6)(3/6)(6/36)$$
$$= P(A)P(B)P(C)$$
Equation 2.5.4:
$$P(B \cap C) = P(\{1, 3), (5,6)\}) = 2/36 \neq (3/6)(6/36)$$
$$= P(B)P(C)$$

2.5.13 11 [= 6 verifications of the form $P(A_i \cap A_j) = P(A_i) \cdot P(A_j)$ + 4 verifications of the form $P(A_i \cap A_j \cap A_k) = P(A_i) \cdot P(A_j) \cdot P(A_k)$ + 1 verification that $P(A_1 \cap A_2 \cap A_3 \cap A_4) = P(A_1) \cdot P(A_2) \cdot P(A_3) \cdot P(A_4)$].

2.5.15 $P(A \cap B \cap C) = 0$ (since the sum of two odd numbers is necessarily even) $\neq P(A) \cdot P(B) \cdot P(C) > 0$, so A, B, and C are not mutually independent. However, $P(A \cap B) = \dfrac{9}{36} = P(A) \cdot P(B) = \dfrac{3}{6} \cdot \dfrac{3}{6}$, $P(A \cap C) = \dfrac{9}{36} = P(A) \cdot P(C) = \dfrac{3}{6} \cdot \dfrac{18}{36}$, and $P(B \cap C) = \dfrac{9}{36} = P(B) \cdot P(C) = \dfrac{3}{6} \cdot \dfrac{18}{36}$, so A, B, and C are pairwise independent.

2.5.17 Let M, L, and G be the events that a student passes the mathematics, language, and general knowledge tests, respectively. Then $P(M) = \dfrac{6175}{9500}$, $P(L) = \dfrac{7600}{9500}$, and $P(G) = \dfrac{8075}{9500}$. P(student fails to qualify) = P(student fails at least one exam) = $1 - P$(student passes all three exams) $= 1 - P(M \cap L \cap G) = 1 - P(M) \cdot P(L) \cdot P(G) = 0.56$.

2.5.19 Let p be the probability of having a winning game card.
Then $0.32 = P$(winning at least once in 5 tries)
$$= 1 - P(\text{not winning in 5 tries})$$
$$= 1 - (1 - p)^5, \text{ so } p = 0.074$$

2.5.21 Andy decides not to shoot at Charley. If he hits him, then Bob, who never misses, would shoot Andy and hit him. So suppose Andy shoots at Bob. The first scenario is that he hits Bob. Then Charley will proceed to shoot at Andy. Andy will shoot back at Charley, and so on, until one of them hits the other. Let CH_i and CM_i denote the events "Charley hits Andy with ith shot" and "Charley misses Andy with ith shot," respectively. Define AH_i and AM_i analogously. Then Andy's chances of survival (given that he has killed Bob) reduce to a countably infinite union of intersections:

$$P(\text{Andy survives}) = \big((CM_1 \cap AH_1) \cup (CM_1 \cap AM_1 \cap CM_2 \cap AH_2$$
$$\cup \, (CM_1 \cap AM_1 \cap CM_2 \cap AM_2 \cap CM_3 \cap AH_3) \cup \cdots\big)$$

Note that each intersection is mutually exclusive of all the others and its component events are independent. Therefore,

$$\begin{aligned} P(\text{Andy survives}) &= P(CM_1)P(AH_1) + P(CM_1)P(AM_1)P(CM_2)P(AH_2) \\ &\quad + P(CM_1)P(AM_1)P(CM_2)P(AM_2)P(CM_3)P(AH_3) + \cdots \\ &= (0.5)(0.3) + (0.5)(0.7)(0.5)(0.3) \\ &\quad + (0.5)(0.7)(0.5)(0.7)(0.5)(0.3) + \cdots \\ &= (0.5)(0.3)\sum_{k=0}^{\infty}(0.35)^k \\ &= (0.15)\left(\frac{1}{1-0.35}\right) \\ &= \frac{3}{13} \end{aligned}$$

Now consider the second scenario. If Andy shoots at Bob and misses, Bob will undoubtedly shoot at (and hit) Charley, since Charley is the more dangerous adversary. Then it will be Andy's turn again. Whether or not he sees another tomorrow will depend on his ability to make that very next shot count. Specifically,

$$P(\text{Andy survives}) = P(\text{Andy hits Bob on second turn})$$
$$= \frac{3}{10}$$

But $\dfrac{3}{10} > \dfrac{3}{13}$, so Andy is better off *not* hitting Bob with his first shot. And because we have already argued it would be foolhardy for Andy to shoot at Charley, Andy's optimal strategy is clear—deliberately miss everyone with the first shot.

2.5.23 Let B be the event that no heads appear, and let A_i be the event that i coins are tossed, $i = 1, 2, \ldots, 6$. Then $P(B) = \sum_{i=1}^{6} P(B|A_i)P(A_i) = \frac{1}{2}\left(\frac{1}{6}\right) + \left(\frac{1}{2}\right)^2\left(\frac{1}{6}\right) + \ldots + \left(\frac{1}{2}\right)^6\left(\frac{1}{6}\right) = \frac{63}{384}$.

2.5.25 P(at least one double six in n throws) $= 1 - P$(no double sixes in n throws) $= 1 - \left(\frac{35}{36}\right)^n$. By trial and error, the smallest n for which P(at least one double six in n throws) exceeds 0.50 is

25 $[1 - \left(\frac{35}{36}\right)^{24} = 0.49; \quad 1 - \left(\frac{35}{36}\right)^{25} = 0.51]$.

2.5.27 Let W, B, and R denote the events of getting a white, black and red chip, respectively, on a given draw. Then P(white appears before red) $= P(W \cup (B \cap W) \cup (B \cap B \cap W) \cup \cdots) =$

$$\frac{w}{w+b+r} + \frac{b}{w+b+r} \cdot \frac{w}{w+b+r} + \left(\frac{b}{w+b+r}\right)^2 \cdot \frac{w}{w+b+r} + \cdots =$$

$$\frac{w}{w+b+r} \cdot \left(\frac{1}{1 - b/(w+b+r)}\right) = \frac{w}{w+r}.$$

2.5.29 P(at least one four) $= 1 - P$(no fours) $= 1 - (0.9)^n$
$1 - (0.9)^n \geq 0.7$ implies $n = 12$

Section 2.6

2.6.1 $2 \cdot 3 \cdot 2 \cdot 2 = 24$

2.6.3 $3 \cdot 3 \cdot 5 = 45$. Included will be aeu and cdx.

2.6.5 There are 9 choices for the first digit (1 through 9), 9 choices for the second digit (0 + whichever eight digits are not appearing in the hundreds place), and 8 choices for the last digit. The number of admissible integers, then, is $9 \cdot 9 \cdot 8 = 648$. For the integer to be odd, the last digit must be either 1, 3, 5, 7, or 9. That leaves 8 choices for the first digit and 8 choices for the second digit, making a total of 320 ($= 8 \cdot 8 \cdot 5$) odd integers.

2.6.7 The bases can be occupied in any of 2^7 ways (each of the seven can be either "empty" or "occupied"). Moreover, the batter can come to the plate facing any of five possible "out" situations (0 through 4). It follows that the number of base-out configurations is $5 \cdot 2^7$, or 640.

2.6.9 $4 \cdot 14 \cdot 6 + 4 \cdot 6 \cdot 5 + 14 \cdot 6 \cdot 5 + 4 \cdot 14 \cdot 5 = 1156$

2.6.11 The number of usable garage codes is $2^8 - 1 = 255$, because the "combination" where none of the buttons is pushed is inadmissable (recall Example 2.6.3). Five additional families can be added before the eight-button system becomes inadequate.

2.6.13 In order to exceed 256, the binary sequence of coins must have a head in the ninth position and at least one head somewhere in the first eight tosses. The number of sequences satisfying those conditions is $2^8 - 1$, or 255. (The "1" corresponds to the sequences TTTTTTTTH, whose value would not exceed 256.)

2.6.15 There are $2 \cdot 3 \cdot 12$ ways if the ace of clubs is not one of the cards and $2 \cdot 1 \cdot 36$ ways if it is. The total is then $2 \cdot 3 \cdot 12 + 2 \cdot 1 \cdot 36 = 144$.

2.6.17 $_6P_3 = 6 \cdot 5 \cdot 4 = 120$

2.6.19 $\log_{10}(30!) \doteq \log_{10}\left(\sqrt{2\pi}\right) + \left(30 + \dfrac{1}{2}\right)\log_{10}(30) - 30\log_{10}e = 32.42246$, which implies that

$30! \doteq 10^{32.42246} = 2.645 \times 10^{32}$.

2.6.21 There are 2 choices for the first digit, 6 choices for the middle digit, and 5 choices for the last digit, so the number of admissible integers that can be formed from the digits 1 through 7 is 60 $(= 2 \cdot 6 \cdot 5)$.

2.6.23 There are 4 different sets of three semesters in which the electives could be taken. For each of those sets, the electives can be selected and arranged in $_{10}P_3$ ways, which means that the number of possible schedules is $4 \cdot {}_{10}P_3$, or 2880.

2.6.25 The number of playing sequences where at least one side is out of order = total number of playing sequences − number of correct playing sequences = $_6P_6 - 1 = 719$.

2.6.27 There are $_2P_2 = 2$ ways for you and a friend to be arranged, $_8P_8$ ways for the other eight to be permuted, and six ways for you and a friend to be in consecutive positions in line. By the multiplication rule, the number of admissible arrangements is $_2P_2 \cdot {}_8P_8 \cdot 6 = 483,840$.

2.6.29 $(13!)^4$

2.6.31 $_9P_2 \cdot {}_4C_1 = 288$

2.6.33 (a) $(4!)(5!) = 2800$
(b) $6(4!)(5!) = 17,280$
(c) $(4!)(5!) = 2880$
(d) $\dbinom{9}{4}(2)(5!) = 30,240$

2.6.35 If the first digit is a 4, the remaining six digits can be arranged in $\dfrac{6!}{3!(1!)^3} = 120$ ways; if the first digit is a 5, the remaining six digits can be arranged in $\dfrac{6!}{2!2!(1!)^2} = 180$ ways. The total number of admissible numbers, then, is $120 + 180 = 300$.

2.6.37 (a) $4! \cdot 3! \cdot 3! = 864$
(b) $3! \cdot 4!3!3! = 5184$ (each of the 3! permutations of the three nationalities can generate $4!3!3!$ arrangements of the ten people in line)
(c) $10! = 3,628,800$
(d) $10!/4!3!3! = 4200$

2.6.39 Imagine a field of 4 entrants (A, B, C, D) assigned to positions 1 through 4, where positions 1 and 2 correspond to the opponents for game 1 and positions 3 and 4 correspond to the opponents for game 2. Although the four players can be assigned to the four positions in 4! ways, not all of those permutations yield different tournaments. For example, $\dfrac{B\ C\ A\ D}{1\ 2\ 3\ 4}$ and $\dfrac{A\ D\ B\ C}{1\ 2\ 3\ 4}$ produce the same set of games, as do $\dfrac{B\ C\ A\ D}{1\ 2\ 3\ 4}$ and $\dfrac{C\ B\ A\ D}{1\ 2\ 3\ 4}$. In general, n games can be arranged in $n!$ ways, and the two players in each game can be permuted in 2! ways. Given a field of $2n$ entrants, then, the number of distinct pairings is $(2n)!/n!(2!)^n$, or $1 \cdot 3 \cdot 5 \cdots (2n-1)$.

2.6.41 The letters in E L E E M O S Y N A R Y minus the pair S Y can be permuted in $10!/3!$ ways. Since S Y can be positioned in front of, within, or behind those ten letters in 11 ways, the number of admissible arrangements is $11 \cdot 10!/3! = 6,652,800$.

2.6.43 Six, because the first four pitches must include two balls and two strikes, which can occur in $4!/2!2! = 6$ ways.

2.6.45 Think of the six points being numbered 1 through 6. Any permutation of three A's and three B's—for example, $\dfrac{A\ A\ B\ B\ A\ B}{1\ 2\ 3\ 4\ 5\ 6}$—corresponds to the three vertices chosen for triangle A and the three for triangle B. It follows that $6!/3!3! = 20$ different sets of two triangles can be drawn.

2.6.47 There are $\dfrac{14!}{2!2!1!2!2!3!1!1!}$ total permutations of the letters. There are $\dfrac{5!}{2!2!1!} = 30$ arrangements of the vowels, only one of which leaves the vowels in their original position. Thus, there are $\dfrac{1}{30} \cdot \dfrac{14!}{2!2!1!2!1!3!1!1!} = 30,270,240$ arrangements of the word leaving the vowels in their original position.

2.6.49 The three courses with A grades can be:
emf, emp, emh, efp, efh, eph, mfp, mfh, mph, fph, or 10 possibilities. From the point of view of Theorem 2.6.2, the grade assignments correspond to the set of permutations of three A's and two B's, which equals $\dfrac{5!}{3!2!} = 10$.

2.6.51 To achieve the two-to-one ratio, six pledges need to be chosen from the set of 10 and three from the set of 15, so the number of admissible classes is $\dbinom{10}{6} \cdot \dbinom{15}{3} = 95,550$.

2.6.53 (a) $\dbinom{9}{4} = 126$

    (b) $\dbinom{5}{2}\dbinom{4}{2} = 60$

    (c) $\dbinom{9}{4} - \dbinom{5}{4} - \dbinom{4}{4} = 120$

2.6.55 Consider a simpler problem: Two teams of two each are to be chosen from a set of four players—A, B, C, and D. Although a single team can be chosen in $\binom{4}{2}$ ways, the number of *pairs* of teams is only $\binom{4}{2} \big/ 2$, because $[(A\,B), (C\,D)]$ and $[(C\,D), (A\,B)]$ would correspond to the same matchup. Applying that reasoning here means that the ten players can split up in $\binom{10}{5} \big/ 2 = 126$ ways.

2.6.57 The four I's need to occupy any of the $\binom{8}{4}$ sets of four spaces between and around the other seven letters. Since the latter can be permuted in $\dfrac{7!}{2!4!1!}$ ways, the total number of admissible arrangements is $\binom{8}{4} \cdot \dfrac{7!}{2!4!1!} = 7350$.

2.6.59 Consider the problem of selecting an unordered sample of n objects from a set of $2n$ objects, where the $2n$ have been divided into two groups, each of size n. Clearly, we could choose n from the first group and 0 from the second group, or $n-1$ from the first group and 1 from the second group, and so on. Altogether, $\binom{2n}{n}$ must equal $\binom{n}{n}\binom{n}{0} + \binom{n}{n-1}\binom{n}{1} + \dots + \binom{n}{0}\binom{n}{n}$. But $\binom{n}{j} = \binom{n}{n-j}$, $j = 0, 1, \dots, n$ so $\binom{2n}{n} = \sum_{j=0}^{n}\binom{n}{j}^2$.

2.6.61 The ratio of two successive terms in the sequence is $\binom{n}{j+1} \big/ \binom{n}{j} = \dfrac{n-j}{j+1}$. For small j, $n - j > j + 1$, implying that the terms are increasing. For $j > \dfrac{n-1}{2}$, though, the ratio is less than 1, meaning the terms are decreasing.

2.6.63 Using Newton's binomial expansion, the equation $(1 + t)^d \cdot (1 + t)^e = (1 + t)^{d+e}$ can be written
$$\left(\sum_{j=0}^{d}\binom{d}{j}t^j\right) \cdot \left(\sum_{j=0}^{e}\binom{e}{j}t^j\right) = \sum_{j=0}^{d+e}\binom{d+e}{j}t^j$$
Since the exponent k can arise as $t^0 \cdot t^k$, $t^1 \cdot t^{k-1}$, $\dots$, or $t^k \cdot t^0$, it follows that
$$\binom{d}{0}\binom{e}{k} + \binom{d}{1}\binom{e}{k-1} + \dots + \binom{d}{k}\binom{e}{0} = \binom{d+e}{k}.$$
That is, $\binom{d+e}{k} = \sum_{j=0}^{k}\binom{d}{j}\binom{e}{k-j}$.

Chapter 2

13

Section 2.7

2.7.1 $\binom{7}{2}\binom{3}{2}/\binom{10}{4}$

2.7.3 P(numbers differ by more than 2) $= 1 - P$(numbers differ by one) $- P$(numbers differ by 2) $= 1$
$- 19/\binom{20}{2} - 18/\binom{20}{2} = \dfrac{153}{190} = 0.81.$

2.7.5 Let A_1 be the event that an urn with $3W$ and $3R$ is sampled; let A_2 be the event that the urn with $5W$ and $1R$ is sampled. Let B be the event that the three chips drawn are white. By Bayes' rule,

$$P(A_2 \,|\, B) = \frac{P(B|A_2)P(A_2)}{P(B|A_1)P(A_1) + P(B|A_2)P(A_2)}$$

$$= \frac{\left[\binom{5}{3}\binom{1}{0}/\binom{6}{3}\right] \cdot (1/10)}{\left[\binom{3}{3}\binom{3}{0}/\binom{6}{3}\right] \cdot (9/10) + \left[\binom{5}{3}\binom{1}{0}/\binom{6}{3}\right] \cdot (1/10)} = \frac{10}{19}$$

2.7.7 $6/6^n = 1/6^{n-1}$

2.7.9 By Theorem, 2.6.2, the $2n$ grains of sand can be arranged in $(2n)!/n!n!$ ways. Two of those arrangements have the property that the colors will completely separate. Therefore, the probability of the latter is $2(n!)^2/(2n)!$.

2.7.11 P(different floors) $= 7!/7^7$; P(same floor) $= 7/7^7 = 1/7^6$. The assumption being made is that all possible departure patterns are equally likely, which is probably not true, since residents living on lower floors would be less inclined to wait for the elevator than would those living on the top floors.

2.7.13 The 10 short pieces and 10 long pieces can be lined up in a row in $20!/(10)!(10)!$ ways. Consider each of the 10 pairs of consecutive pieces as defining the reconstructed sticks. Each of those pairs could combine a short piece (S) and a long piece (L) in two ways: SL or LS. Therefore, the number of permutations that would produce 10 sticks, each having a short and a long component is 2^{10}, so the desired probability is $2^{10}/\binom{20}{10}$.

2.7.15 Any of $\binom{k}{2}$ people could share any of 365 possible birthdays. The remaining $k - 2$ people can generate $364 \cdot 363 \cdots (365 - k + 2)$ sequences of distinct birthdays. Therefore, P(exactly one match) $= \binom{k}{2} \cdot 365 \cdot 364 \cdots (365 - k + 2)/365^k.$

2.7.17 To get a flush, Dana needs to draw any three of the remaining eleven diamonds. Since only forty-seven cards are effectively left in the deck (others may already have been dealt, but their identities are unknown), $P(\text{Dana draws to flush}) = \binom{11}{3} \Big/ \binom{47}{3}$.

2.7.19 There are two pairs of cards that would give Tim a straight flush (5 of clubs and 7 of clubs or 7 of clubs and 10 of clubs). Therefore, $P(\text{Tim draws to straight flush}) = 2 \Big/ \binom{47}{2}$. A flush, by definition, consists of five cards in the same suit whose denominations are not all consecutive. It follows that $P(\text{Tim draws to flush}) = \left[\binom{10}{2} - 2\right] \Big/ \binom{47}{2}$, where the "2" refers to the straight flushes cited earlier.

2.7.21 $\binom{5}{3}\binom{4}{2}^3\binom{3}{1}\binom{4}{2}\binom{2}{1}\binom{4}{1} \Big/ \binom{52}{9}$

2.7.23 $\left[\binom{2}{1}\binom{2}{1}\right]^4 \binom{32}{4} \Big/ \binom{48}{12}$

Chapter 3

Section 3.2

3.2.1 The number of days, k, the stock rises is binomial with $n = 4$ and $p = 0.25$. The stock will be the same after four days if $k = 2$. The probability that $k = 2$ is $\binom{4}{2}(0.25)^2(0.75)^2 = 0.211$

3.2.3 The probability of k sightings is given by the binomial probability model with $n = 10,000$ and $p = 1/100,000$. The probability of at least one genuine sighting is the probability that $k \geq 1$. The probability of the complementary event, $k = 0$, is $(99,999/100,000)^{10,000} = 0.905$. Thus, the probability that $k \geq 1$ is $1 - 0.905 = 0.095$.

3.2.5 The probability the day's work will get done = the probability that 3 or fewer are out of service $= \sum_{k=0}^{3}\binom{10}{k}(0.05)^k(0.95)^{10-k} = 0.599 + 0.315 + 0.075 + 0.010 = 0.999$

3.2.7 The number of 6's obtained in n tosses is binomial with $p = 1/6$. The first probability in question has $n = 6$. The probability that $k \geq 1$ is $1 - (5/6)^6 = 1 - 0.33 = 0.67$.
For the second situation, $n = 12$. The probability that $k \geq 2$ one minus the probability that $k = 0$ or 1, which is
$$1 - (5/6)^{12} = 12(1/6)(5/6)^{11} = 0.62.$$
Finally, take $n = 18$. The probability that $k \geq 3$ is one minus the probability that $k = 0, 1,$ or 2, which is
$$1 - (5/6)^{18} - 18(1/6)(5/6)^{17} - 153(1/6)^2(5/6)^{16} = 0.60.$$

3.2.9 The number of girls is binomial with $n = 4$ and $p = 1/2$. The probability of two girls and two boys is $\binom{4}{2}(0.4)^4 = 0.375$. The probability of three and one is $2\binom{4}{3}(0.5)^4 = 0.5$, so the latter is more likely.

3.2.11 The probability it takes k calls to get four drivers is $\binom{k-1}{3}0.80^4 0.20^{k-4}$. We seek the smallest number n so that $\sum_{k=4}^{n}\binom{k-1}{3}0.80^4 0.20^{k-4} \geq 0.95$. By trial and error, $n = 7$.

3.2.13 (1) The probability that any one of the seven measurements will be in the interval $(1/2, 1)$ is 0.50. The probability that exactly three will fall in the interval is $\binom{7}{3}0.5^7$
$= 0.273$
(2) The probability that any one of the seven measurements will be in the interval $(3/4, 1)$ is 0.25. The probability that fewer than 3 will fall in the interval is
$$\sum_{k=0}^{2}\binom{7}{k}(0.25)^k(0.75)^{7-k} = 0.756$$

3.2.15 By the binomial theorem $(x+y)^n = \sum_{k=0}^{n} \binom{n}{k} x^k y^{n-k}$. Let $x = p$ and $y = 1-p$.

Then $1 = [p + (1-p)]^n = \sum_{k=0}^{n} \binom{n}{k} p^k (1-p)^{n-k}$

3.2.17 In the notation of Question 3.2.16, $p_1 \doteq 0.5$ and $p_2 = 0.3$, with $n = 10$.
Then the probability of 3 of Outcome 1 and 5 of Outcome 2 is

$$\frac{10!}{3!5!2!}(0.5)^3(0.3)^5(0.2)^2 = 0.031$$

3.2.19 "At least twice as many black bears as tan-colored" translates into spotting 4, 5, or 6 black bears. The probability is

$$\frac{\binom{6}{4}\binom{3}{2}}{\binom{9}{6}} + \frac{\binom{6}{5}\binom{3}{1}}{\binom{9}{6}} + \frac{\binom{6}{6}\binom{3}{0}}{\binom{9}{6}} = 64/84$$

3.2.21 The probability that k nuclear missiles will be destroyed by the anti-ballistic missiles is hypergeometric with $N = 10$, $n = 7$, $r = 6$, and $w = 10 - 6 = 4$. The probability the Country B will be hit by at least one nuclear missile is one minus the probability that $k = 6$, or

$$1 - \frac{\binom{6}{6}\binom{4}{1}}{\binom{10}{7}} = 0.967$$

3.2.23 The probabilities for the number of men chosen are hypergeometric with $N = 18$, $n = 5$, $r = 8$, and $w = 10$. The event that both men and women are represented is the complement of the event that 0 or 5 men will be chosen, or

$$1 - \frac{\binom{8}{0}\binom{10}{5}}{\binom{18}{5}} + \frac{\binom{8}{5}\binom{10}{0}}{\binom{18}{5}} = 1 - \frac{252}{8568} + \frac{56}{8568} = 0.964$$

3.2.25 First, calculate the probability that exactly one real diamond is taken during the first three grabs. There are three possible positions in the sequence for the real diamond, so this probability is

$$\frac{3(10)(25)(24)}{(35)(34)(33)}$$

The probability of a real diamond being taken on the fourth removal is 9/32. Thus, the desired probability is

$$\frac{3(10)(25)(24)}{(35)(34)(33)} \times \frac{9}{32} = \frac{162,000}{1,256,640} = 0.129$$

3.2.27 The k-th term of $(1 + \mu)^N = \binom{N}{k}\mu^k$

$$(1 + \mu)^r (1 + \mu)^{N-r} = \left(\sum_{i=1}^{r} \binom{r}{i}\mu^i \right) \left(\sum_{j=1}^{N-r} \binom{N-r}{j}\mu^j \right)$$

The k-th term of this product is $\sum_{i=1}^{r} \binom{r}{i}\binom{N-r}{k-i}\mu^k$

Equating coefficients gives $\binom{N}{k} = \sum_{i=1}^{k} \binom{r}{i}\binom{N-r}{k-i}$.

Dividing through by $\binom{N}{k}$ shows that the hypergeometric terms sum to 1.

3.2.29 For any value of r = number of defective items, the probability of accepting the sample is

$$p_r = \frac{\binom{r}{0}\binom{100-r}{10}}{\binom{100}{10}} + \frac{\binom{r}{1}\binom{100-r}{9}}{\binom{100}{10}}$$

Then the operating characteristic curve is the plot of the presumed percent defective versus the probability of accepting the shipment, or $100(r/100) = r$ on the x-axis and p_r on the y-axis. If there are 16 defective, you will accept the shipment approximately 50% of the time.

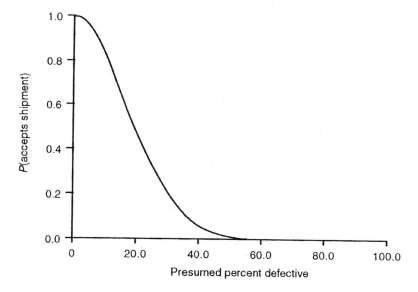

3.2.31 First, calculate the probability that the first group contains two disease carriers and the others have one each. The probability of this, according to Question 3.2.30, is

$$\frac{\binom{7}{2}\binom{7}{1}\binom{7}{1}}{\binom{21}{4}} = 49/285$$

The probability that either of the other two groups has 2 carriers and the others have one is the same. Thus, the probability that each group has at least one diseased member is

$$3\frac{49}{285} = \frac{49}{95} = 0.516$$

Then the probability that at least one group is disease free is $1 - 0.516 = 0.484$.

3.2.33 In the notation of Question 3.2.32, let $n_1 = 5$, $n_2 = 4$, $n_3 = 4$, $n_4 = 3$, so $N = 16$. The sample size is given to be $n = 8$, and $k_1 = k_2 = k_3 = k_4 = 2$. Then the probability that each class has two representatives is

$$\frac{\binom{5}{2}\binom{4}{2}\binom{4}{2}\binom{3}{2}}{\binom{16}{8}} = \frac{(10)(6)(6)(3)}{12,870} = \frac{1080}{12,870} = 0.084$$

Section 3.3

3.3.1 (a) Each outcome has probability 1/10

Outcome	X = larger no. drawn
1, 2	2
1, 3	3
1, 4	4
1, 5	5
2, 3	3
2, 4	4
2, 5	5
3, 4	4
3, 5	5
4, 5	5

Counting the number of each value of the larger of the two and multiplying by 1/10 gives the pdf:

k	$p_X(k)$
2	1/10
3	2/10
4	3/10
5	4/10

(b)

Outcome	X = larger no. drawn	V = sum of two nos.
1, 2	2	3
1, 3	3	4
1, 4	4	5
1, 5	5	6
2, 3	3	5
2, 4	4	6
2, 5	5	7
3, 4	4	7
3, 5	5	8
4, 5	5	9

k	$p_X(k)$
3	1/10
4	1/10
5	2/10
6	2/10
7	2/10
8	1/10
9	1/10

3.3.3 $p_X(k) = P(X = k) = P(X \le k) - P(X \le k - 1)$. But the event $(X \le k)$ occurs when all three dice are $\le k$ and that can occur in k^3 ways. Thus $P(X = k) = k^3/216$.
Similarly, $P(X \le k - 1) = (k - 1)^3/216$. Thus $p_X(k) = k^3/216 - (k - 1)^3/216$.

3.3.5

Outcomes	V = no. heads $-$ no. tails
(H, H, H)	3
(H, H, T) (H, T, H) (T, H, H)	1
(T, T, H) (T, H, T) (T, H, H)	-1
(T, T, T)	-3

$p_X(3) = 1/8$, $p_X(1) = 3/8$, $p_X(-1) = 3/8$, $p_X(-3) = 1/8$

3.3.7 This is similar to Question 3.3.5. If there are k steps to the right (heads), then there are $4 - k$ steps to the left (tails). The final position X is number of heads $-$ number of tails $= k - (4 - k) = 2k - 4$.

The probability of this is the binomial of getting k heads in 4 tosses $= \binom{4}{k}\dfrac{1}{16}$.

Thus, $p_X(2k - 4) = \binom{4}{k}\dfrac{1}{16}$, $k = 0, 1, 2, 3, 4$

3.3.9 Let us consider the case $k = 0$ as an example. If you are on the left, with your friend on your immediate right, you two can stand in positions 1, 2, 3, or 4. The remaining people can stand in 3! ways. Each of these must be multiplied by 2, since your friend could be the one on the left. The total number of permutations of the five people is 5!
Thus,

$$p_X(0) = (2)(4)(3!)/5! = 48/120 = 4/10$$

In a similar manner

$$p_X(1) = (2)(3)(3!)/5! = 36/120 = 3/10$$
$$p_X(2) = (2)(2)(3!)/5! = 24/120 = 2/10$$
$$p_X(3) = (2)(1)(3!)/5! = 12/120 = 1/10$$

3.3.11 By Theorem 3.3.1, $p_{2X+1}(k) = p_X\left(\dfrac{k-1}{2}\right) = \binom{4}{\frac{k-1}{2}}\left(\dfrac{2}{3}\right)^{\frac{k-1}{2}}\left(\dfrac{1}{3}\right)^{4-\frac{k-1}{2}}$, $k = 1, 3, 5, 7, 9$

3.3.13 $F_X(k) = P(X \le k) = \displaystyle\sum_{j=0}^{k} P(X = j) = \sum_{j=0}^{k}\binom{4}{k}\left(\dfrac{1}{6}\right)^k\left(\dfrac{5}{6}\right)^{4-k}$

3.3.15 See the solution to Question 3.3.3.

Chapter 3

21

Section 3.4

3.4.1 $P(0 \le Y \le 1/2) = \int_0^{1/2} 4y^3 dy = y^4 \Big|_0^{1/2} = 1/16$

3.4.3 $P\left(|Y - 1/2| < 1/4\right) = P(1/4 < Y < 3/4) = \int_{1/4}^{3/4} \frac{3}{2} y^2 \, dy = \frac{y^3}{2} \Big|_{1/4}^{3/4} = \frac{27}{128} - \frac{1}{128} = \frac{26}{128} = \frac{13}{64}$

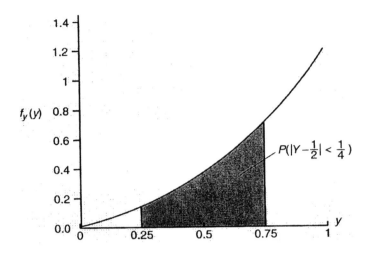

3.4.5 (a) $\int_{10}^{\infty} 0.2 e^{-0.2y} dy = -e^{-0.2y} \Big|_{10}^{\infty} = e^{-2} = 0.135$

(b) If A = probability customer leaves on first trip, and B = probability customer leaves on second trip, then $P(A) = P(B) = 0.135$. In this notation,

$$p_X(1) = P(A)P(B^C) + P(A^C)P(B) = 2(0.865)(0.135) = 0.23355$$

3.4.7 $F_Y(y) = P(Y \le y) = \int_0^y 4t^3 dt = t^4 \Big|_0^y = y^4$. Then $P(Y \le 1/2) = F_Y(1/2) = (1/2)^4 = 1/16$

3.4.9 For $y < -1$, $F_Y(y) = 0$

For $-1 \le y < 0$, $F_Y(y) = \int_{-1}^{y}(1+t)dt = \frac{1}{2} + y + \frac{1}{2}y^2$

For $0 \le y \le 1$, $F_Y(y) = \int_{-1}^{y}(1-|t|)dt = \frac{1}{2} + \int_{0}^{y}(1-t)dt = \frac{1}{2} + y - \frac{1}{2}y^2$

For $y > 1$, $F_Y(y) = 1$

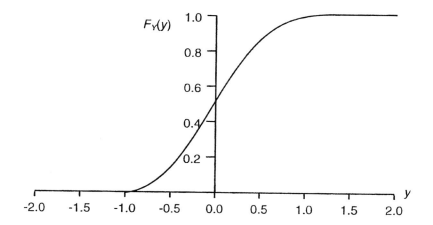

3.4.11 (a) $P(Y < 2) = F_Y(2)$, since F_Y is continuous over [0, 2]. Then $F_Y(2) = \ln 2 = 0.693$

(b) $P(2 < Y \le 2.5) = F_Y(2.5) - F_Y(2) = \ln 2.5 - \ln 2 = 0.223$

(c) The probability is the same as (b) since F_Y is continuous over $[0, e]$

(d) $f_Y(y) = \frac{d}{dy}F_Y(y) = \frac{d}{dy}\ln y = \frac{1}{y}$, $1 \le y \le e$

3.4.13 $f_Y(y) = \frac{d}{dy}\frac{1}{12}(y^2 + y^3) = \frac{1}{6}y + \frac{1}{4}y^2, 0 \le y \le 2$

3.4.15 $f_W(w) = \frac{1}{3}f_Y\left(\frac{w-2}{3}\right) = \frac{1}{3}\frac{3}{2}\left(\frac{w-2}{3}\right)^2 = \frac{1}{18}(w^2 - 4w + 4)$, $-1 \le w \le 5$

3.4.17 $F_Y(y) = \int_{0}^{y}(1/\lambda)e^{-t/\lambda}dt = 1 - e^{-t/\lambda}$, so

$h(y) = \frac{(1/\lambda)e^{-y/\lambda}}{1 - (1 - e^{-y/\lambda})} = 1/\lambda$

Since the hazard rate is constant, the item does not age. Its reliability does not decrease over time.

Section 3.5

3.5.1 $E(X) = -1(0.935) + 2(0.0514) + 18(0.0115) + 180(0.0016) + 1,300(1.35 \times 10^{-4})$
$+ 2,600(6.12 \times 10^{-6}) + 10,000(1.12 \times 10^{-7}) = -0.144668$

3.5.3 $E(X) = \$30,000(0.857375) + \$18,000(0.135375) + \$6,000(0.007125) + (-\$6,000)(0.000125)$
$= \$28,200.00$

3.5.5 The average number of customers identified is $1000\left(\dfrac{5000}{60,000}\right) = 250/3$

Now let n be the number of calls necessary to find an average of 100 new customers. Then

$n\left(\dfrac{5000}{60,000}\right) = 100$ implies $n = 1200$.

3.5.7 This is a hypergeometric problem where r = number of students needing vaccinations = 125 and w = number of students already vaccinated = $642 - 125 = 517$. An absenteeism rate of 12% corresponds to a sample $n = (0.12)(642) \doteq 77$ missing students. The expected number of unvaccinated students who are absent when the physician visits is $\dfrac{125(77)}{125 + 517} \doteq 15$.

3.5.9 $E(Y) = \displaystyle\int_0^3 y\left(\frac{1}{9}y^2\right)dy = \frac{1}{9}\int_0^3 y^3 dy = \left.\frac{y^4}{36}\right|_0^3 = \frac{9}{4}$ years

3.5.11 $E(Y) = \displaystyle\int_0^\infty y \cdot \lambda e^{-\lambda y}dy = \left.\left(-ye^{-\lambda y} - \frac{1}{\lambda}e^{-\lambda y}\right)\right|_0^\infty = \frac{1}{\lambda}$

3.5.13 Let X be the number of cars passing the emissions test. Then X is binomial with $n = 200$ and $p = 0.80$. Two formulas for $E(X)$ are:

(1) $E(X) = \displaystyle\sum_{k=1}^n k\binom{n}{k}p^k(1-p)^{n-k} = \sum_{k=1}^{200} k\binom{200}{k}(0.80)^k(0.20)^{200-k}$

(2) $E(X) = np = 200(0.80) = 160$

3.5.15 If birthdays are randomly distributed throughout the year, the city should expect revenue of $(\$50)(74,806)(30/365)$ or $\$307,421.92$.

3.5.17 For the experiment described, construct the table:

Sample	Larger of the two, k
1, 2	2
1, 3	3
1, 4	4
2, 3	3
2, 4	4
3, 4	4

Each of the six samples is equally likely to be drawn, so $p_X(2) = 1/6$, $p_X(3) = 2/6$, and $p_X(4) = 3/6$. Then $E(X) = 2(1/6) + 3(2/6) + 4(3/6) = 20/6 = 10/3$.

3.5.19 The "fair" ante is the expected value of X, which is

$$\sum_{k=1}^{9} 2^k \left(\frac{1}{2^k}\right) + \sum_{k=10}^{\infty} 1000 \left(\frac{1}{2^k}\right) = 9 + \frac{1000}{2^{10}} \sum_{k=0}^{\infty} \left(\frac{1}{2^k}\right)$$

$$= 9 + \frac{1000}{2^{10}} \frac{1}{1 - \frac{1}{2}} = 9 + \frac{1000}{512} = \frac{5608}{512} = \$10.95$$

3.5.21 $p_X(1) = \dfrac{6}{216} = \dfrac{1}{36}$

$p_X(2) = \dfrac{3(6)(5)}{216} = \dfrac{15}{36}$

$p_X(3) = \dfrac{6(5)(4)}{216} = \dfrac{20}{36}$

$E(X) = 1 \cdot \dfrac{1}{36} + 2 \cdot \dfrac{15}{36} + 3 \cdot \dfrac{20}{36} = \dfrac{91}{36}$

3.5.23 Let X be the length of the series. Then

$$p_X(k) = 2\binom{k-1}{3}\left(\frac{1}{2}\right)^k, \ k = 4, 5, 6, 7$$

$$E(X) = \sum_{k=4}^{7}(k)(2)\binom{k-1}{3}\left(\frac{1}{2}\right)^k$$

$$= 4\left(\frac{2}{16}\right) + 5\left(\frac{4}{16}\right) + 6\left(\frac{5}{16}\right) + 7\left(\frac{5}{16}\right) = \frac{93}{16} = 5.8125$$

3.5.25 $E(X) = \displaystyle\sum_{k=1}^{r} k \frac{\dbinom{r}{k}\dbinom{w}{n-k}}{\dbinom{r+w}{n}} = \sum_{k=1}^{r} k \frac{\dfrac{r!}{k!(r-k)!}\dbinom{w}{n-k}}{\dfrac{(r+w)!}{n!(r+w-n)!}}$

Factor out the presumed value of $E(X) = rn/(r+w)$:

$E(X) = \dfrac{rn}{r+w} \displaystyle\sum_{k=1}^{r} \frac{\dfrac{(r-1)!}{(k-1)!(r-k)!}\dbinom{w}{n-k}}{\dfrac{(r-1+w)!}{(n-1)!(r+w-n)!}} = \dfrac{rn}{r+w} \sum_{k=1}^{r} \frac{\dbinom{r-1}{k-1}\dbinom{w}{n-k}}{\dbinom{r-1+w}{n-1}}$

Change the index of summation to begin at 0, which gives

$E(X) = \dfrac{rn}{r+w} \displaystyle\sum_{k=0}^{r-1} \frac{\dbinom{r-1}{k}\dbinom{w}{n-1-k}}{\dbinom{r-1+w}{n-1}}$. The terms of the summation are urn probabilities where

there are $r-1$ red balls, w white balls, and a sample size of $n-1$ is drawn. Since these are the probabilities of a hypergeometric pdf, the sum is one. This leaves us with the desired

equality $E(X) = \dfrac{rn}{r+w}$.

3.5.27 $E(3X - 4) = 3E(X) - 4 = 3(10)(2/5) - 4 = 8$

3.5.29 (1) First find $f_Y(y)$: $F_Y(y) = P(Y \le y) = P(X^3 \le y) = P(X \le y^{1/3}) = F_X(y^{1/3})$.

Then $f_Y(y) = \dfrac{1}{3}y^{-2/3} f_X(y^{1/3}) = \dfrac{2}{3}(y^{-2/3} - y^{-1/3})$

$E(Y) = \displaystyle\int_0^1 y \frac{2}{3}(y^{-2/3} - y^{-1/3})dy = \frac{2}{3}\int_0^1 (y^{1/3} - y^{2/3})dy$

$= \dfrac{2}{3}\left[\left(\dfrac{3}{4}y^{4/3} - \dfrac{3}{5}y^{5/3}\right)\right]_0^1 = \dfrac{1}{10}$

(2) $E(Y) = \displaystyle\int_0^1 x^3 2(1-x)dx = 2\int_0^1 (x^3 - x^4)dx$

$= 2\left[\dfrac{x^4}{4} - \dfrac{x^5}{5}\right]_0^1 = \dfrac{1}{10}$

3.5.31 $E(\text{Volume}) = \displaystyle\int_0^\infty 5y^2 6y(1-y)dy = 30\int_0^1 (y^3 - y^4)dy = 30\left[\dfrac{1}{4}y^4 - \dfrac{1}{5}y^5\right]_0^1 = 1.5 \text{ in}^3$

3.5.33 For the graph pictured to be a pdf, $t = 4$ and $f_Y(y) = y/8$.

$E(Y^2) = \displaystyle\int_0^4 y^2 \frac{y}{8}dy = \frac{1}{8}\int_0^4 y^3 dy = \left.\dfrac{y^4}{32}\right|_0^4 = 8$

Chapter 3

3.5.35 $1 = \sum_{i=1}^{n} ki = k\frac{n(n+1)}{2}$ implies $k = \dfrac{2}{n(n+1)}$

$E\left(\dfrac{1}{X}\right) = \sum_{i=1}^{n}\dfrac{1}{i}\dfrac{2}{n(n+1)}i = 2/(n+1)$

Section 3.6

3.6.1 If sampling is done with replacement, X is binomial with $n = 2$ and $p = 2/5$. By Theorem 3.5.1, $\mu = 2(2/5) = 4/5$.

$E(X^2) = 0 \cdot (9/25) + 1 \cdot (12/25) + 4 \cdot (4/25) = 28/25$. Then $\mathrm{Var}(X) = 28/25 - (4/5)^2 = 12/25$.

3.6.3 Since X is hypergeometric, $\mu = \dfrac{3(6)}{10} = \dfrac{9}{5}$

$E(X^2) = \sum_{k=0}^{3} k^2 \dfrac{\binom{6}{k}\binom{4}{3-k}}{\binom{10}{3}} = 0 \cdot (4/120) + 1 \cdot (36/120) + 4 \cdot (60/12) + 9 \cdot (20/120) =$

$456/120 = 38/10$
$\mathrm{Var}(X) = 38/10 - (9/5)^2 = 28/50 = 0.56$, and $\sigma = 0.748$

3.6.5 $\mu = \int_0^1 y3(1-y)^2 dy = 3\int_0^1 (y - 2y^2 + y^3)dy = 1/4$

$E(Y^2) = \int_0^1 y^2 3(1-y)^2 dy = 3\int_0^1 (y^2 - 2y^3 + y^4)dy = 1/10$

$\mathrm{Var}(Y) = 1/10 - (1/4)^2 = 3/80$

3.6.7 $f_Y(y) = \begin{cases} 1-y, & 0 \le y \le 1 \\ 1/2, & 2 \le y \le 3 \\ 0, & \text{elsewhere} \end{cases}$

$\mu = \int_0^1 y(1-y)dy + \int_2^3 y\left(\dfrac{1}{2}\right)dy = 17/12$

$E(Y^2) = \int_0^1 y^2(1-y)dy + \int_2^3 y^2\left(\dfrac{1}{2}\right)dy = 13/4$

$\sigma = \sqrt{13/4 - (17/12)^2} = \sqrt{179}/12 = 1.115$

3.6.9 Let Y = Frankie's selection. Johnny wants to choose k so that $E[(Y-k)^2]$ is minimized. The minimum occurs when $k = E(Y) = (a+b)/2$ (see Question 3.6.13).

3.6.11 Using integration by parts, we find that

$$E(Y^2) = \int_0^\infty y^2 \lambda e^{-\lambda y}\,dy = -y^2\,e^{-\lambda y}\Big|_0^\infty + \int_0^\infty 2y e^{-\lambda y}\,dy = 0 + \int_0^\infty 2y e^{-\lambda y}\,dy,$$

The right hand term is $2\int_0^\infty y e^{-\lambda y}\,dy = \dfrac{2}{\lambda}\int_0^\infty y\lambda e^{-\lambda y}\,dy = \dfrac{2}{\lambda}E(Y) = \dfrac{2}{\lambda}\dfrac{1}{\lambda} = \dfrac{2}{\lambda^2}.$

Then $\mathrm{Var}(Y) = E(Y^2) - E(Y)^2 = \dfrac{2}{\lambda^2} - \left(\dfrac{1}{\lambda}\right)^2 = \dfrac{1}{\lambda^2}.$

3.6.13 $E[(X-a)^2] = E[((X-\mu)+(\mu-a))^2]$
$= E[(X-\mu)^2] + E[(\mu-a)^2] + 2(\mu-a)E(X-\mu)$
$= \mathrm{Var}(X) + (\mu-a)^2$, since $E(X-\mu) = 0$. This is minimized when $a = \mu$, so the minimum of $g(a) = \mathrm{Var}(X)$.

3.6.15 $\mathrm{Var}\left(\dfrac{5}{9}(Y-32)\right) = \left(\dfrac{5}{9}\right)^2 \mathrm{Var}(Y)$, by Theorem 3.6.2. So $\sigma\left(\dfrac{5}{9}(Y-32)\right) = \left(\dfrac{5}{9}\right)\sigma(Y) = \dfrac{5}{9}(15.7)$
$= 8.7°C.$

3.6.17 $f_Y(y) = \dfrac{1}{b-a}f_U\left(\dfrac{y-a}{b-a}\right) = \dfrac{1}{b-a}$, $(b-a)(0)+a \le y \le (b-a)(1)+a$, or

$f_Y(y) = \dfrac{1}{b-a}$, $a \le y \le b$, which is the uniform pdf over $[a, b]$

(b) $\mathrm{Var}(Y) = \mathrm{Var}[(b-a)U + a]\ (b-a)^2\,\mathrm{Var}(U) = (b-a)^2/12$

3.6.19 For the given f_Y, $\mu = 1$ and $\sigma = 1$.
$$\gamma_1 = \dfrac{E\left[(Y-1)^3\right]}{1} = \sum_{j=0}^{3}\binom{3}{j}E(Y^j)(-1)^{3-j} = \sum_{j=0}^{3}\binom{3}{j}(j!)(-1)^{3-j}$$
$(1)(1)(-1) + (3)(1)(1) + (3)(2)(-1) + (1)(6)(1) = 2$

3.6.21 $10 = E\left[(W-2)^3\right] = \sum_{j=0}^{3}\binom{3}{j}E(W^j)(-2)^{3-j}$
$= (1)(1)(-8) + (3)(2)(4) + (3)E(W^2)(-2) + (1)(4)(1)$
This would imply that $E(W^2) = 5/3$. In that case, $\mathrm{Var}(W) = 5/3 - (2)^2 < 0$, which is not possible.

3.6.23 (a) Question 3.4.6 established that Y is a pdf for any positive integer n. As a corollary, we know that $1 = \int_0^1 (n+2)(n+1)y^n(1-y)dy$ or equivalently, for any positive

integer n, $\int_0^1 y^n(1-y)dy = \dfrac{1}{(n+2)(n+1)}$

Then $E(Y^2) =$

$\int_0^1 y^n(n+2)(n+1)y^n(1-y)dy = (n+2)(n+1)\int_0^1 y^{n+2}(1-y)dy = \dfrac{(n+2)(n+1)}{(n+4)(n+3)}$

By a similar argument, $E(Y) = \dfrac{(n+2)(n+1)}{(n+4)(n+3)} = \dfrac{(n+1)}{(n+3)}$

Thus, $\text{Var}(Y) = E(Y^2) - E(Y)^2 = \dfrac{(n+2)(n+1)}{(n+4)(n+3)} - \dfrac{(n+1)^2}{(n+3)^2} = \dfrac{2(n+1)}{(n+4)(n+3)^2}$

(b) Then $E(Y^k) =$

$\int_0^1 y^k(n+2)(n+1)y^n(1-y)dy = (n+2)(n+1)\int_0^1 y^{n+k}(1-y)dy = \dfrac{(n+2)(n+1)}{(n+k+2)(n+k+1)}$

Section 3.7

3.7.1 $1 = \sum_{x,y} p(x,y) = c\sum_{x,y} xy =$

$c[(1)(1) + (2)(1) + (2)(2) + (3)(1)] = 10c$, so $c = 1/10$

3.7.3 $1 = \int_0^1 \int_0^y c(x+y)dxdy = c\int_0^1 \left[\dfrac{x^2}{2} + xy\right]_0^y dy$

$= c\int_0^1 \dfrac{3y^2}{2}dy = c\left[\dfrac{y^3}{2}\right]_0^1 = \dfrac{c}{2}$, so $c = 2$.

3.7.5 $P(X=x, Y=y) = \dfrac{\binom{3}{x}\binom{2}{y}\binom{4}{3-x-y}}{\binom{9}{3}}$, $0 \le x \le 3$, $0 \le y \le 2$, $x + y \le 3$

3.7.7 $P(X > Y) = p_{X,Y}(1, 0) + p_{X,Y}(2, 0) + p_{X,Y}(2, 1)$
$= 6/50 + 4/50 + 3/50 = 13/50$

3.7.9

		Number of 2's, X		
		0	1	2
Number	0	16/36	8/36	1/36
of 3's, Y	1	8/36	2/36	0
	2	1/36	0	0

From the matrix above, we calculate

$p_Z(0) = p_{X,Y}(0, 0) = 16/36$
$p_Z(1) = p_{X,Y}(0, 1) + p_{X,Y}(1, 0) = 2(8/36) = 16/36$
$p_Z(2) = p_{X,Y}(0, 2) + p_{X,Y}(2, 0) + p_{X,Y}(1, 1) = 4/36$

3.7.11 $P(Y < 3X) = \int_{-\infty}^{\infty} \int_x^{3x} 2e^{-(x+y)} dy dx = \int_{-\infty}^{\infty} e^{-x} \left(\int_x^{3x} 2e^{-y} dy \right) dx$

$$= 2 \int_{-\infty}^{\infty} e^{-x} \left(\left[-e^{-y} \right]_x^{3x} \right) dx = 2 \int_{-\infty}^{\infty} e^{-x} \left[e^{-x} - e^{-3x} \right] dx$$

$$= 2 \int_0^{\infty} \left[e^{-2x} - e^{-4x} \right] dx = 2 \left[-\frac{1}{2} e^{-2x} + \frac{1}{4} e^{-4x} \right]_0^{\infty} = \frac{1}{2}$$

3.7.13 $P(X < 2Y) = \int_0^1 \int_{x/2}^1 (x + y) dy dx$

$$= \int_0^1 \int_{x/2}^1 x \, dy dx + \int_0^1 \int_{x/2}^1 y \, dy dx$$

$$= \int_0^1 \left[x - \frac{x^2}{2} \right] dx + \int_0^1 \left[\frac{1}{2} - \frac{x^2}{8} \right] dx$$

$$= \left[\frac{x^2}{2} - \frac{x^3}{6} \right]_0^1 + \left[\frac{x}{2} - \frac{x^3}{24} \right]_0^1 = \frac{19}{24}$$

3.7.15 The set where $y > h/2$ is a triangle with height $h/2$ and base $b/2$. Its area is $bh/8$. Thus the area of the set where $y < h/2$ is $bh/2 - bh/8 = 3bh/8$. The probability that a randomly chosen point will fall in the lower half of the triangle is $(3bh/8)/(bh/2) = 3/4$.

3.7.17 From the solution to Question 3.7.8, $p_X(x) = 1/8 + 2/8 + 1/8 = 1/2$, $x = 0, 1$, $p_Y(0) = 1/8$, $p_Y(1) = 3/8$, $p_Y(2) = 3/8$, $p_Y(3) = 1/8$.

3.7.19 (a) $f_X(x) = \int_0^1 \frac{1}{2} dy = \frac{y}{2} \Big|_0^1 = \frac{1}{2}$, $0 \le x \le 2$

$f_Y(y) = \int_0^2 \frac{1}{2} dx = \frac{x}{2} \Big|_0^2 = 1, 0 \le y \le 1$

Chapter 3

(b) $f_X(x) = \int_0^1 \frac{3}{2} y^2 \, dy = \frac{1}{2} y^3 \Big|_0^1 = 1, \, 0 \leq x \leq 2$

$f_Y(y) = \int_0^2 \frac{3}{2} y^2 \, dx = \frac{3}{2} y^2 x \Big|_0^2 = 3y^2, \, 0 \leq y \leq 1$

(c) $f_X(x) = \int_0^1 \frac{2}{3}(x+2y) dy = \frac{2}{3}\left(xy + y^2\right) \Big|_0^1 = \frac{2}{3}(x+1), \, 0 \leq x \leq 1$

$f_Y(y) = \int_0^1 \frac{2}{3}(x+2y) dx = \frac{2}{3}\left(\frac{x^2}{2} + 2xy\right) \Big|_0^1 = \frac{4}{3} y + \frac{1}{3}, \, 0 \leq y \leq 1$

(d) $f_X(x) = c \int_0^1 (x+y) dy = c\left(xy + \frac{y^2}{2}\right)\Big|_0^1 = c\left(x + \frac{1}{2}\right), \, 0 \leq x \leq 1$

In order for the above to be a density,

$1 = \int_0^1 c\left(x + \frac{1}{2}\right) dx = c\left(\frac{x^2}{2} + \frac{x}{2}\right)\Big|_0^1 = c$, so

$f_X(x) = x + \frac{1}{2}, \, 0 \leq x \leq 1$

$f_Y(y) = y + \frac{1}{2}, \, 0 \leq y \leq 1$, by symmetry of the joint pdf

(e) $f_X(x) = \int_0^1 4xy \, dy = 2xy^2 \Big|_0^1 = 2x, \, 0 \leq x \leq 1$

$f_Y(y) = 2y, \, 0 \leq y \leq 1$, by the symmetry of the joint pdf

(f) $f_X(x) = \int_0^\infty xye^{-(x+y)} dy = xe^{-x} \int_0^\infty ye^{-y} dy$

$= xe^{-x}(-ye^{-y} - e^{-y})\Big|_0^\infty = xe^{-x}, \, 0 \leq x$

$f_Y(y) = ye^{-y}, \, 0 \leq y$, by symmetry of the joint pdf

(g) $f_X(x) = \int_0^\infty ye^{-xy-y} dy = \int_0^\infty ye^{-(x+1)y} dy$

Integrating by parts gives

$\left(-\frac{y}{x+1}e^{-(x+1)y} - \left(\frac{1}{x+1}\right)^2 e^{-(x+1)y}\right)\Big|_0^\infty = \left(\frac{1}{x+1}\right)^2, \, 0 < x$

$f_Y(y) = \int_0^\infty ye^{-xy-y} dx = \int_0^\infty ye^{-y}e^{-xy} dx = ye^{-y}\left(-\frac{1}{y}\right)e^{-xy}\Big|_0^\infty = e^{-y},$

where $0 \leq y$.

3.7.21 $f_X(x) = \int_0^{1-x} 6(1 - x - y)\,dy = 6\left(y - xy - \dfrac{y^2}{2}\right)\Bigg|_0^{1-x}$

$$= 6\left[(1 - x) - x(1 - x) - \dfrac{(1-x)^2}{2}\right] = 3 - 6x + 3x^2,\ 0 \le x \le 1$$

3.7.23 $p_X(x) = \displaystyle\sum_{y=0}^{4-x} \dfrac{4!}{x!\,y!\,(4 - x - y)!}\left(\dfrac{1}{2}\right)^x \left(\dfrac{1}{3}\right)^y \left(\dfrac{1}{6}\right)^{4-x-y}$

$$= \dfrac{4!}{x!(4-x)!}\left(\dfrac{1}{2}\right)^x \sum_{y=0}^{4-x} \dfrac{(4-x)!}{y!\,[(4-x)-y]!}\left(\dfrac{1}{3}\right)^y \left(\dfrac{1}{6}\right)^{4-x-y} = \dfrac{4!}{x!(4-x)!}\left(\dfrac{1}{2}\right)^x \left(\dfrac{1}{3} + \dfrac{1}{6}\right)^{4-x}$$

$$= \binom{4}{x}\left(\dfrac{1}{2}\right)^x \left(\dfrac{1}{2}\right)^{4-x}$$

Thus, X is binomial with $n = 4$ and $p = 1/2$. Similarly, Y is binomial with $n = 4$ and $p = 1/3$.

3.7.25 (a) $S = \{(H, 1), (H, 2), (H, 3), (H, 4), (H, 5), (H, 6), (T, 1), (T, 2), (T, 3), (T, 4), (T, 5),$
$(T, 6)\}$

   (b) $F_{X,Y}(1, 2) = P(X \le 1, Y \le 2)$
$= P(\{(H, 1), (H, 2), (T, 1), (T, 2)\}) = 4/12 = 1/3$

3.7.27 (a) $F_{X,Y}(u, v) = \displaystyle\int_0^u \int_0^v \dfrac{3}{2} y^2\,dy\,dx = \int_0^u \left[\dfrac{1}{2} y^3\right]_0^v dx = \int_0^u \dfrac{1}{2} v^3\,dx = \dfrac{1}{2}uv^3$

   (b) $F_{X,Y}(u, v) = \displaystyle\int_0^u \int_0^v \dfrac{2}{3}(x + 2y)\,dy\,dx = \int_0^u \left[\dfrac{2}{3}(xy + y^2)\right]_0^v dx = \int_0^u \dfrac{2}{3}(vx + v^2\,dx) = \dfrac{1}{3}u^2 v + \dfrac{2}{3}uv^2$

   (c) $F_{X,Y}(u, v) = \displaystyle\int_0^u \int_0^v 4xy\,dy\,dx = \int_0^u x\left[2y^2\right|_0^v\right] dx = 2v^2 \int_0^u x\,dx = u^2 v^2$

3.7.29 By Theorem 3.7.3, $f_{X,Y} = \dfrac{\partial^2}{\partial x \partial y} F_{X,Y} = \dfrac{\partial^2}{\partial x \partial y}(xy) = \dfrac{\partial}{\partial x}\left(\dfrac{\partial}{\partial y} xy\right)$

$$= \dfrac{\partial}{\partial x}(x) = 1,\ 0 \le x \le 1,\ 0 \le y \le 1.$$

The graph of $f_{X,Y}$ is a plane of height one over the unit square.

3.7.31 First note that $1 = F_{X,Y}(1, 1) = k[4(1^2)(1^2) + 5(1)(1^4)] = 9k$, so $k = 1/9$.

$$\text{Then } f_{X,Y} = \frac{\partial^2}{\partial x \partial y} F_{X,Y} = \frac{\partial^2}{\partial x \partial y}\left(\frac{4}{9}x^2 y^2 + \frac{5}{9}xy^4\right)$$

$$= \frac{\partial}{\partial x}\frac{\partial}{\partial y}\left(\frac{4}{9}x^2 y^2 + \frac{5}{9}xy^4\right) = \frac{\partial}{\partial x}\left(\frac{8}{9}x^2 y + \frac{20}{9}xy^3\right) = \frac{16}{9}xy + \frac{20}{9}y^3$$

$$P(0 < X < 1/2,\ 1/2 < Y < 1) = \int_0^{1/2}\int_{1/2}^{1}\left(\frac{16}{9}xy + \frac{20}{9}y^3\right)dy\,dx$$

$$= \int_0^{1/2}\frac{8}{9}xy^2 + \frac{5}{9}y^4\Big|_{1/2}^{1}\,dx = \int_0^{1/2}\left(\frac{2}{3}x + \frac{25}{48}\right)dx$$

$$= \frac{1}{3}x^2 + \frac{25}{48}x\ \Big|_0^{1/2} = 11/32$$

3.7.33 $P(X_1 \geq 1050, X_2 \geq 1050, X_3 \geq 1050, X_4 \geq 1050)$

$$= \int_{1050}^{\infty}\int_{1050}^{\infty}\int_{1050}^{\infty}\int_{1050}^{\infty}\prod_{i=1}^{4}\frac{1}{1000}e^{-x_i/1000}dx_1 dx_2 dx_3 dx_4$$

$$= \left(\int_{1050}^{\infty}\frac{1}{1000}e^{-x/1000}dx\right)^4 = (e^{1.05})^4 = 0.015$$

3.7.35 $p_{X,Y}(0, 1) = \sum_{z=0}^{2}p_{X,Y,Z}(0,1,z)$

$$= \frac{3!}{0!\,1!}\left(\frac{1}{2}\right)^0\left(\frac{1}{12}\right)^1\sum_{z=0}^{2}\frac{1}{z!(2-z)!}\left(\frac{1}{6}\right)^z\left(\frac{1}{4}\right)^{2-z}$$

$$= \frac{3!}{0!\,1!}\left(\frac{1}{2}\right)^0\left(\frac{1}{12}\right)^1\left(\frac{1}{2}\right)\sum_{z=0}^{2}\frac{2!}{z!(2-z)!}\left(\frac{1}{6}\right)^z\left(\frac{1}{4}\right)^{2-z}$$

$$= \frac{3!}{0!\,1!}\left(\frac{1}{2}\right)^0\left(\frac{1}{12}\right)^1\left(\frac{1}{2}\right)\left(\frac{1}{6} + \frac{1}{4}\right)^2 = \frac{25}{576}$$

3.7.37 $f_{W,X}(w, x) = \int_0^1\int_0^1 f_{W,X,Y,Z}(w,x,y,z)dy\,dz = \int_0^1\int_0^1 16wxyz\ dy\,dz$

$$= \int_0^1\left[8wxy^2 z\right]_0^1 dz = \int_0^1[8wxz]dz = \left[4wxz^2\right]_0^1 = 4wx,\ 0 < w,\ x < 1$$

$$P(0 < W < 1/2,\ 1/2 < X < 1) = \int_0^{1/2}\int_{1/2}^{1}4wx\ dx\,dw$$

$$= \int_0^{1/2}2w\left[x^2\right]_{1/2}^1 dx = \int_0^{1/2}\frac{3}{2}w\ dw = \frac{3}{4}w^2\Big|_0^{1/2} = \frac{3}{16}$$

3.7.39 The marginal pdfs for $f_{X,Y}$ are $f_X(x) = \lambda e^{-\lambda x}$ and $f_Y(y) = \lambda e^{-\lambda y}$ (Hint: see the solution to 3.7.19(f)). Their product is $f_{X,Y}$, so X and Y are independent. The probability that one component fails to last 1000 hours is $1 - e^{-1000\lambda}$. Because of independence of the two components, the probability that two components both fail is the square of that, or $(1 - e^{-1000\lambda})^2$.

3.7.41 First, note $k = 2$. Then, 2 times area of $A = P(Y \geq 3/4)$. Also, 2 times area of $B = P(X \geq 3/4)$. The square C is the set $(X \geq 3/4) \cap (Y \geq 3/4)$. However, C is in the region where the density is 0. Thus, $P((X \geq 3/4) \cap (Y \geq 3/4))$ is zero, but the product $P(X \geq 3/4)P(Y \geq 3/4)$ is not zero.

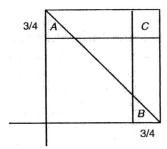

3.7.43 $P(Y < X) = \int_0^1 \int_0^x f_{X,Y}(x,y)\, dy dx = \int_0^1 \int_0^x (2x)(3y^2)\, dy dx = \int_0^1 2x^4\, dx = \dfrac{2}{5}$

3.7.45 $P\left(\dfrac{Y}{X} > 2\right) = P(Y > 2X) = \int_0^1 \int_0^{y/2} (2x)(1) dx dy = \int_0^1 [x^2]_0^{y/2}\, dy = \dfrac{y^3}{12}\Big|_0^1 = \dfrac{1}{12}$

3.7.47 Take $a = c = 0$, $b = d = 1/2$. Then
$$P(0 < X < 1/2, 0 < Y < 1/2) = \int_0^{1/2} \int_0^{1/2} (2x + y - 2xy) dy dx = 5/32.$$

$f_X(x) = \int_0^1 (2x + y - 2xy) dy = x + 1/2$, so $P(0 < X < 1/2) = \int_0^{1/2} \left(x + \dfrac{1}{2}\right) dx = \dfrac{3}{8}.$

$f_Y(y) = \int_0^1 (2x + y - 2xy) dx = 1$, so $P(0 < X < 1/2) = 1/2$. But, $5/32 \neq (3/8)(1/2)$

3.7.49 Let K be the region of the plane where $f_{X,Y} \neq 0$. If K is not a rectangle with sides parallel to the coordinate axes, there exists a rectangle
$$A = \{(x, y)\,|\; a \leq x \leq b, c \leq y \leq d\}$$
with $A \cap K = \emptyset$, but for $A_1 = \{(x, y)\,|\, a \leq x \leq b, \text{all } y\}$ and $A_2 = \{(x, y)\,|\; \text{all } x, c \leq y \leq d\}$, $A_1 \cap K \neq \emptyset$ and $A_2 \cap K \neq \emptyset$. Then $P(A) = 0$, but $P(A_1) \neq 0$ and $P(A_2) \neq 0$. However, $A = A_1 \cap A_2$, so $P(A_1 \cap A_2) \neq P(A_1)P(A_2)$.

3.7.51 (a) $P(X_1 < 1/2) = \int_0^{1/2} 4x^3 dx = x^4 \Big|_0^{1/2} = 1/16$

(b) This asks for the probability of exactly one success in a binomial experiment with $n = 4$ and $p = 1/16$, so the probability is $\dbinom{4}{1}(1/16)^1(15/16)^3 = 0.206$.

(c) $f_{X_1,X_2,X_3,X_4}(x_1,x_2,x_3,x_4) = \prod_{j=1}^{4} 4x_j^3 = 256(x_1x_2x_3x_4)^3$

(d) $F_{X_2,X_3}(x_2,x_3) = \int_0^{x_3}\int_0^{x_2}(4s^3)(4t^3)dsdt = \int_0^{x_2}4s^3ds\int_0^{x_3}4t^3dt = x_2^4x_3^4,\ 0 \le x_2, x_3 \le 1.$

Section 3.8

3.8.1 (a) $p_{X+Y}(w) = \sum_{\text{all }x}p_X(x)p_Y(w-x)$. Since $p_X(x) = 0$ for negative x, we can take the lower limit of the sum to be 0. Since $p_Y(w-x) = 0$ for $w - x < 0$, or $x > w$, we can take the upper limit of the sum to be w. Then we obtain

$$p_{X+Y}(w) = \sum_{k=0}^{w} \pm e^{-\lambda}\frac{\lambda^k}{k}e^{-\mu}\frac{\mu^{w-k}}{(w-k)!} = e^{-(\lambda+\mu)}\sum_{k=0}^{w}\frac{1}{k!(w-k)!}\lambda^k\mu^{w-k}$$

$$= e^{-(\lambda+\mu)}\frac{1}{w!}\sum_{k=0}^{w}\frac{w!}{k!(w-k)!}\lambda^k\mu^{w-k} = e^{-(\lambda+\mu)}\frac{1}{w!}(\lambda+\mu)^w,$$

$w = 0, 1, 2, \ldots$

This pdf has the same form as the ones for X and Y, but with parameter $\lambda + \mu$.

(b) $p_{X+Y}(w) = \sum_{\text{all }x}p_X(x)p_Y(w-x)$. The lower limit of the sum is 1. For this pdf, we must have $w - k \ge 1$ so the upper limit of the sum is $w - 1$. Then $p_{X+Y}(w) =$

$$\sum_{k=1}^{w-1}(1-p)^{k-1}p(1-p)^{w-k-1}p = (1-p)^{w-2}p^2\sum_{k=1}^{w-1}1 = (w-1)(1-p)^{w-2}p^2,$$

$w = 2, 3, 4, \ldots$

The pdf for $X + Y$ does not have the same form as those for X and Y, but Section 4.5 will show that they all belong to the same family—the negative binomial.

3.8.3 First suppose that $0 \le w \le 1$. As in the previous problem the upper limit of the integral is w, and $f_{X+Y}(w) = \int_0^w (1)(1)dx = w$.

Now consider the case $1 \le w \le 2$. Here, the first integrand vanishes unless x is ≤ 1. Also, the second pdf is 0 unless $w - x \le 1$ or $x \ge w - 1$. Then $f_{X+Y}(w) = \int_{w-1}^1 (1)(1)dx = 2 - w$.

In summary, $f_{X+Y}(w) = \begin{cases} w & 0 \le w \le 1 \\ 2 - w & 1 \le w \le 2 \end{cases}$

3.8.5 $F_W(w) = P(W \le w) = P(Y^2 \le w) = P\left(Y \le \sqrt{w}\right) = f = F_Y\left(\sqrt{w}\right)$

Now differentiate both sides to obtain

$$f_W(w) = \frac{d}{dw}F_W(w) = \frac{d}{dw}F_Y\left(\sqrt{w}\right) = \frac{1}{2\sqrt{w}}f_Y\left(\sqrt{w}\right) = \frac{1}{2\sqrt{w}}$$

where $0 \le w \le 1$.

3.8.7 (a) Let $V = XY$. Then

$$f_V(v) = \int_{-\infty}^{\infty} \frac{1}{|x|} f_X(v/x) f_Y(x) dx$$

Since $f_X(v/x) \neq 0$ when $0 \leq v/x \leq 1$, then we need only consider $v \leq x$.

Similarly, $f_Y(x) \neq 0$ implies $x \leq 1$. Thus the integral becomes $\int_v^1 \frac{1}{x} dx = \ln x \Big|_v^1 = -\ln v$,

$0 \leq v \leq 1$

(b) Again let $V = XY$. Since the range of integration here is the same as in Part (a), we can write

$$f_V(v) = \int_{-\infty}^{\infty} \frac{1}{|x|} f_X(v/x) f_Y(x) dx = \int_v^1 \frac{1}{x} 2(v/x)(2x) dx = 4v \int_v^1 \frac{1}{x} dx = 4v(1 - \ln v), \ 0 \leq v \leq 1$$

3.8.9 Let $W = Y/X$. Then

$$f_W(w) = \int_{-\infty}^{\infty} |x| f_X(x) f_Y(wx) dx = \int_0^{\infty} x(xe^{-x}) e^{-wx} dx = \int_0^{\infty} x^2 e^{-(1+w)x} dx$$

$$= \frac{1}{1+w} \left(\int_0^{\infty} x^2 (1+w) e^{-(1+w)x} dx \right)$$

Let V be the exponential random variable with parameter $1 + w$. Then the quantity in parentheses above is $E(V^2)$.

But $E(V)^2 = \text{Var}(V) + E(V^2) = \frac{1}{(1+w)^2} + \frac{1}{(1+w)^2} = \frac{2}{(1+w)^2}$ (See Question 3.6.11)

Thus, $f_W(w) = \frac{1}{1+w} \left(\frac{2}{(1+w)^2} \right) = \frac{2}{(1+w)^3}, \ 0 \leq w$

Section 3.9

3.9.1 Let X_i be the number from the i-th draw, $i = 1, \ldots, r$. Then for each i,

$E(X_i) = \frac{1+2+\ldots+n}{n} = \frac{n+1}{2}$. The sum of the numbers drawn is $\sum_{i=1}^{r} X_i$, so the expected value

of the sum is $\sum_{i=1}^{r} E(X_i) = \frac{r(n+1)}{2}$.

3.9.3 From Question 3.7.19(c), we have $f_X(x) = \frac{2}{3}(x+1)$, $0 \leq x \leq 1$, so

$$E(X) = \int_0^1 x \frac{2}{3}(x+1) dx = \frac{2}{3} \int_0^1 (x^2 + x) dx = \frac{5}{9}$$

Also, $f_Y(y) = \frac{4}{3} y + \frac{1}{3}$, $0 \leq y \leq 1$, so

$$E(Y) = \int_0^1 y \left(\frac{4}{3} y + \frac{1}{3} \right) dy = \int_0^1 \left(\frac{4}{3} y^2 + \frac{1}{3} y \right) dy = \frac{11}{8}$$

Then $E(X + Y) = E(X) + E(Y) = \frac{5}{9} + \frac{11}{18} = \frac{7}{6}$

3.9.5 $\mu = E\left(\sum_{i=1}^{n} a_i X_i\right) = \sum_{i=1}^{n} a_i E(X_i) = \sum_{i=1}^{n} a_i \mu = \mu \sum_{i=1}^{n} a_i$, so the given equality occurs if and only if

$$\sum_{i=1}^{n} a_i = 1.$$

3.9.7 (a) $E(X_i)$ is the probability that the i-th ball drawn is red, $1 \leq i \leq n$. Draw the balls in order without replacement, but do not note the colors. Then look at the i-th ball <u>first</u>. The probability that it is red is surely independent of when it is drawn. Thus, all of these expected values are the same and each equals $r/(r + w)$.

(b) Let X be the number of red balls drawn. Then $X = \sum_{i=1}^{n} X_i$ and

$$E(X) = \sum_{i=1}^{n} E(X_i) = nr/(r + w)$$

3.9.9 First note that $1 = \int_{10}^{20} \int_{0}^{20} k(x + y)dydx = 3000$, so $k = \dfrac{1}{3000}$.

If $\dfrac{1}{R} = \dfrac{1}{X} + \dfrac{1}{Y}$, then $R = \dfrac{XY}{X + Y}$

$E(R) = \dfrac{1}{3000} \int_{10}^{20} \int_{10}^{20} \dfrac{xy}{x + y}(x + y)dydx = \dfrac{1}{3000} \int_{10}^{20} \int_{10}^{20} xydydx = 7.5.$

3.9.11 The area of the triangle is the random variable $W = \dfrac{1}{2} XY$. Then

$$E\left(\frac{1}{2} XY\right) = \frac{1}{2} E(XY) = \frac{1}{2} E(X)E(Y) = \frac{1}{2} \cdot \frac{1}{2} \cdot \frac{1}{2} = \frac{1}{8}$$

3.9.13 The random variables are independent and have the same exponential pdf, so $\text{Var}(X + Y) = \text{Var}(X) + \text{Var}(Y)$. By Question 3.6.11, $\text{Var}(X) = \text{Var}(Y) = \dfrac{1}{\lambda^2}$, so $\text{Var}(X + Y) = \dfrac{2}{\lambda^2}$.

3.9.15 First note that $E\left[\left(\sqrt{Y_1 Y_2}\right)^2\right] = E[Y_1 Y_2] = E(Y_1)E(Y_2) = \left(\dfrac{1}{2}\right)\left(\dfrac{1}{2}\right) = \dfrac{1}{4}$, since the Y_i are independent, for $i = 1, 2$.

$E\left(\sqrt{Y_1 Y_2}\right) = E\left(\sqrt{Y_1}\right)E\left(\sqrt{Y_2}\right) = \dfrac{2}{3} \cdot \dfrac{2}{3} = \dfrac{4}{9}$, since $E(Y_i) = \int_{0}^{1} \sqrt{y} \cdot 1\, dy = \dfrac{2}{3}$, $i = 1, 2$.

Then $\text{Var}\left(\sqrt{Y_1 Y_2}\right) = \dfrac{1}{4} - \left(\dfrac{4}{9}\right)^2 = \dfrac{17}{324}$

3.9.17 Let U_i be the number of calls during the i-th hour in the normal nine hour work day. Then $U = U_1 + U_2 + \ldots + U_9$ is the number of calls during this nine hour period. $E(U) = 9(7) = 63$. For a Poisson random variable, the variance is equal to the mean, so $\text{Var}(U) = 9(7) = 63$. Similarly, if V is the number of calls during the off hours, $E(V) = \text{Var}(V) = 15(4) = 60$. Let the total cost be the random variable $W = 50U + 60V$. Then $E(W) = E(50U + 60V) = 50E(U) + 60E(V) = 50(63) + 60(60) = 6750$; $\text{Var}(W) = \text{Var}(50U + 60V) = 50^2\text{Var}(U) + 60^2\text{Var}(V) = 50^2(63) + 60^2(60) = 373,500$.

3.9.19 Let R_i be the resistance of the i-th resistor, $1 \le i \le 6$. Assume the R_i are independent and each has standard deviation σ. Then the variance of the circuit resistance is $\text{Var}\left(\sum_{i=1}^{6} R_i \right) = 6\sigma^2$. The circuit must have $6\sigma^2 \le (0.4)^2$ or $\sigma \le 0.163$.

3.9.21 $\text{Var}\left(\dfrac{2D}{T^2} \right) = 4\text{Var}\left(\dfrac{D}{T^2} \right) = 4\left(\dfrac{1}{\mu_T^2} \right)^2 \sigma_D^2 + 4\left(\dfrac{-2\mu_D}{\mu_T^3} \right)\sigma_T^2$

$= \dfrac{4}{\mu_T^4}\left(\sigma_D^2 + \dfrac{4\mu_D^2}{\mu_T^2}\sigma_T^2 \right) = \dfrac{4}{\mu_T^4}\left((0.0025)^2 + \dfrac{4\mu_D^2}{\mu_T^2}(0.045)^2 \right)$

For the first set up, this last expression becomes

$\dfrac{4}{(0.5)^4}\left((0.0025)^2 + \dfrac{4(4^2)}{(0.5)^2}(0.045)^2 \right) = 33.178$

For the second set up, the approximation is

$\dfrac{4}{(1)^4}\left((0.0025)^2 + \dfrac{4(16^2)}{(1)^2}(0.045)^2 \right) = 8.294$

3.9.23 $\sigma_A^2 = \left(\dfrac{1}{2}\mu_h \right)^2 (\sigma_a^2 + \sigma_b^2) + \left[\dfrac{1}{2}(\mu_a + \mu_b) \right]^2 \sigma_h^2$, so

$\sigma_A = \dfrac{1}{2}\sqrt{\mu_h^2\sigma_a^2 + \mu_h^2\sigma_b^2 + (\mu_a + \mu_b)^2\sigma_h^2}$

Section 3.10

3.10.1 $P(Y_3' < 5) = \int_0^5 f_{Y_3'}(y)\,dy$

$= \int_0^5 \dfrac{4!}{(3-1)!(4-3)!10}\dfrac{y}{10}^{3-1}\left(1 - \dfrac{y}{10}\right)^{4-3}\dfrac{1}{10}\,dy$

$= \dfrac{12}{10^4}\int_0^5 y^2(10 - y)\,dy = \dfrac{12}{10^4}\left[\dfrac{10}{3}y^3 - \dfrac{1}{4}y^4 \right]_0^5$

$= \dfrac{12}{10^4}\left[\dfrac{10}{3}5^3 - \dfrac{1}{4}5^4 \right] = 5/16$

3.10.3 $P(Y_2' > y_{60}) = 1 - P(Y_2' < y_{60}) = 1 - P(Y_1 < y_{60}, Y_2 < y_{60})$
$$= 1 - P(Y_1 < y_{60})P(Y_2 < y_{60}) = 1 - (0.60)(0.60) = 0.64$$

3.10.5 $P(Y_1' > m) = P(Y_1, \ldots, Y_n > m) = \left(\dfrac{1}{2}\right)^n$

$P(Y_n' > m) = 1 - P(Y_n' < m) = 1 - P(Y_1, \ldots, Y_n < m)$

$$= 1 - P(Y_1 < m) \cdot \ldots \cdot P(Y_n < m) = 1 - \left(\dfrac{1}{2}\right)^n$$

If $n \geq 2$, the latter probability is greater.

3.10.7 $P(0.6 < Y_4' < 0.7) = F_{Y_4'}(0.7) - F_{Y_4'}(0.6)$

$$= \int_{0.6}^{0.7} \dfrac{6!}{(4-1)!\,(6-4)!} y^{4-1}(1-y)^{6-4}(1)dy \quad \text{(by Theorem 3.10.1)}$$

$$= \int_{0.6}^{0.7} 60y^3(1-y)^2\,dy = \int_{0.6}^{0.7} 60(y^3 - 2y^4 + y^5)dy$$

$$= (15y^4 - 24y^5 + 10y^6)\Big|_{0.6}^{0.7} = 0.74431 - 0.54432 = 0.19999$$

3.10.9 $P(Y_{\min} > 20) = P(Y_1 > 20, Y_2 > 20, \ldots, Y_n > 20) = P(Y_1 > 20)P(Y_2 > 20) \ldots P(Y_n > 20)$
$= [P(Y > 20)]^n$. But 20 is the median of Y, so $P(Y > 20) = 1/2$. Thus, $P(Y_{\min} > 20) = (1/2)^n$.

3.10.11 The graphed pdf is the function $f_Y(y) = 2y$, so $F_Y(y) = y^2$
Then $f_{Y_4'}(y) = 20y^6(1-y^2)2y = 40y^7(1-y^2)$ and $F_{Y_4'}(y) = 5y^8 - 4y^{10}$.

$P(Y_4' > 0.75) = 1 - F_{Y_4'}(0.75) = 1 - 0.275 = 0.725$

The probability that none of the schools will have fewer than 10% of their students bused is

$$P(Y_{\min} > 0.1) = 1 - F_{Y_{\min}}(0.1) = 1 - \int_0^{0.1} 10y(1-y^2)^4\,dy$$

$$= 1 - \left[-(1-y^2)^5\right]_0^{0.1} = 0.951 \quad \text{(see Question 3.10.8)}$$

3.10.13 If $Y_1, Y_2, \ldots Y_n$ is a random sample from the uniform distribution over $[0, 1]$, then by Theorem
3.10.1, the quantity

$$\dfrac{n!}{(i-1)!(n-i)!}\left[F_Y(y)\right]^{i-1}\left[1 - F_Y(y)\right]^{n-i} f_y(y) = \dfrac{n!}{(i-1)!(n-i)!} y^{i-1}(1-y)^{n-i}$$

is the pdf of the i-th order statistic. Thus,

$$1 = \int_0^1 \dfrac{n!}{(i-1)!(n-i)!} y^{i-1}(1-y)^{n-i}\,dy = \dfrac{n!}{(i-1)!(n-i)!} \int_0^1 y^{i-1}(1-y)^{n-i}\,dy$$

or, equivalently,

$$\int_0^1 y^{i-1}(1-y)^{n-i}\,dy = \dfrac{(i-1)!(n-i)!}{n!}$$

3.10.15 This question translates to asking for the probability that a random sample of three
independent uniform random variables on $[0, 1]$ has range $R \leq 1/2$. Example 3.10.6
establishes that $F_R(r) = 3r^2 - 2r^3$. The desired probability is $F_R(1/2) = 3(1/2)^2 - 2(1/2)^3 = 0.5$.

Section 3.11

3.11.1 $p_X(x) = \dfrac{x+1+x\cdot 1}{21} + \dfrac{x+2+x\cdot 2}{21} = \dfrac{3+5x}{21}, x = 1, 2$

$p_{Y|x}(y) = \dfrac{p_{X,Y}(x,y)}{p_X(x)} = \dfrac{(x+y+xy)}{3+5x}, y = 1, 2$

3.11.3 $p_{Y|x}(y) = \dfrac{p_{X,Y}(x,y)}{p_X(x)} = \dfrac{\dbinom{8}{x}\dbinom{6}{y}\dbinom{4}{3-x-y}}{\dbinom{18}{3}} \div \dfrac{\dbinom{8}{x}\dbinom{10}{3-x}}{\dbinom{18}{3}} = \dfrac{\dbinom{6}{y}\dbinom{4}{3-x-y}}{\dbinom{10}{3-x}}$, with $0 \le y \le 3-x$

3.11.5 (a) $1/k = \displaystyle\sum_{x=1}^{3}\sum_{y=1}^{3}(x+y) = 36$, so $k = 1/36$

(b) $p_X(x) = \dfrac{1}{36}\displaystyle\sum_{y=1}^{3}(x+y) = \dfrac{1}{36}(3x+6)$

$p_{Y|x}(1) = \dfrac{p_{X,Y}(x,1)}{p_X(x)} = \dfrac{\dfrac{1}{36}(x+1)}{\dfrac{1}{36}(3x+6)} = \dfrac{x+1}{3x+6}, x = 1, 2, 3$

3.11.7 $p_Z(z) = \dfrac{1\cdot 1 + 1\cdot z + 1\cdot z}{54} + \dfrac{1\cdot 2 + 1\cdot z + 2\cdot z}{54} + \dfrac{2\cdot 1 + 2\cdot z + 1\cdot z}{54} + \dfrac{2\cdot 2 + 2\cdot z + 2\cdot z}{54}$

$= \dfrac{9+12z}{54}, z = 1, 2$

Then $p_{X,Y|z}(x,y) = \dfrac{xy + xz + yz}{9+12z}, x = 1, 2 \quad y = 1, 2 \quad z = 1, 2$

3.11.9 $p_{X|x+y=n}(x) = \dfrac{P(X=k, X+Y=n)}{P(X+Y=n)} = \dfrac{P(X=k, Y=n-k)}{P(X+Y=n)}$

$= \dfrac{e^{-\lambda}\dfrac{\lambda^k}{k!} e^{-\mu}\dfrac{\mu^{n-k}}{(n-k)!}}{e^{-(\lambda+\mu)}\dfrac{(\lambda+\mu)^n}{n!}} = \dfrac{n!}{k!(n-k)!}\left(\dfrac{\lambda}{\lambda+\mu}\right)^k\left(\dfrac{\mu}{\lambda+\mu}\right)^{n-k}$

but the right hand term is a binomial probability with parameters n and $\lambda/(\lambda + \mu)$.

3.11.11 $P(X > s + t \mid X > t) =$

$$\frac{P(X > s+t \text{ and } X > t)}{P(X > t)} = \frac{P(X > s+t)}{P(X > t)}$$

$$= \frac{(1/\lambda)\int_{s+t}^{\infty} e^{-x/\lambda}dx}{(1/\lambda)\int_{t}^{\infty} e^{-x/\lambda}dx} = \frac{-(1/\lambda)e^{-x/\lambda}\Big|_{s+t}^{\infty}}{-(1/\lambda)e^{-x/\lambda}\Big|_{t}^{\infty}} = \frac{(1/\lambda)e^{-(s+t)/\lambda}}{(1/\lambda)e^{-t/\lambda}} =$$

$$e^{-s/\lambda} = \int_{s}^{\infty} (1/\lambda)e^{-x/\lambda}dx = P(X > s)$$

3.11.13 $f_X(x) = \int_0^1 (x+y)dy = xy + \dfrac{y^2}{2}\Big|_0^1 = x + \dfrac{1}{2}, \ 0 \le x \le 1$

$$f_{Y|x}(y) = \frac{f_{X,Y}(x,y)}{f_X(x)} = \frac{x+y}{x+\dfrac{1}{2}}, \ 0 \le y \le 1$$

3.11.15 $f_{X,Y}(x,y) = f_{Y|x}(y)f_X(x) = \left(\dfrac{2y+4x}{1+4x}\right)\dfrac{1}{3}(1+4x) = \dfrac{1}{3}(2y+4x)$

$$f_Y(y) = \int_0^1 \frac{1}{3}(2y+4x)dx = \frac{1}{3}(2xy + 2x^2)\Big|_0^1 = \frac{1}{3}(2y+2), \text{ with } 0 \le y \le 1$$

3.11.17 $f_Y(y) = \int_0^y 2\,dx = 2y$

$$f_{X|y}(x) = \frac{2}{2y} = \frac{1}{y}, \ 0 < x < y$$

$$f_{X|\frac{3}{4}}(x) = \frac{1}{\dfrac{3}{4}} = \frac{4}{3}, \ 0 < x < 3/4$$

$$P(0 < X < 1/2 \mid Y = 3/4) = \int_0^{1/2} \frac{4}{3}dx = \frac{2}{3}$$

3.11.19 $f_{X_4,X_5}(x_4,x_5) = \int_0^1\int_0^1\int_0^1 32x_1x_2x_3x_4x_5\, dx_1dx_2dx_3 = 4x_4x_5, \ 0 < x_4, x_5 < 1$

$$f_{X_1,X_2,X_3|x_4,x_5}(x_1,x_2,x_3) = \frac{32x_1x_2x_3x_4x_5}{4x_4x_5} = 8x_1x_2x_3, \qquad 0 < x_1, x_2, x_3 < 1$$

Note: the five random variables are independent, so the conditional pdfs are just the marginal pdfs.

Chapter 3

Section 3.12

3.12.1 Let X be a random variable with $p_X(k) = 1/n$, for $k = 0, 1, 2, \ldots, n-1$, and 0 otherwise.

$$M_X(t) = E(E^{tX}) = \sum_{k=0}^{n-1} e^{tk} p_X(k) = \sum_{k=0}^{n-1} e^{tk} \frac{1}{n} = \frac{1}{n} \sum_{k=0}^{n-1} (e^t)^k = \frac{1 - e^{nt}}{n(1 - e^t)}.$$

(Recall that $1 + r + \ldots + r^{n-1} = \dfrac{1 - r^n}{1 - r}$).

3.12.3 For the given binomial random variable,

$$E(e^{tX}) = M_X(t) = \left(1 - \frac{1}{3} + \frac{1}{3}e^t\right)^{10}.$$ Set $t = 3$ to obtain $E(e^{3X}) = \frac{1}{3^{10}}(2 + e^3)^{10}.$

3.12.5 (a) Normal with $\mu = 0$ and $\sigma^2 = 12$
  (b) Exponential with $\lambda = 2$
  (c) Binomial with $n = 4$ and $p = 1/2$
  (d) Geometric with $p = 0.3$

3.12.7 $M_X(t) = E(e^{tX}) = \sum_{k=0}^{\infty} e^{tk} e^{-\lambda} \dfrac{\lambda^k}{k!} = e^{-\lambda} \sum_{k=0}^{\infty} \dfrac{(\lambda e^t)^k}{k!} = e^{\lambda(e^t - 1)}$

3.12.9 $M_Y^{(1)}(t) = \dfrac{d}{dt} e^{t^2/2} = t e^{t^2/2}$

$M_Y^{(2)}(t) = \dfrac{d}{dt} t e^{t^2/2} = t(t e^{t^2/2}) + e^{t^2/2} = (t^2 + 1) e^{t^2/2}$

$M_Y^{(3)}(t) = \dfrac{d}{dt}(t^2 + 1)e^{t^2/2} = (t^2 + 1)t e^{t^2/2} + 2t e^{t^2/2}$ and $E(Y^3) = M_Y^{(3)}(0) = 0$

3.12.11 $M_Y^{(1)}(t) = \dfrac{d}{dt} e^{at + b^2 t^2/2} = (a + b^2 t)e^{at + b^2 t^2/2}$, so $M_Y^{(1)}(0) = a$

$M_Y^{(2)}(t) = (a + b^2 t)^2 e^{at + b^2 t^2/2} + b^2 e^{at + b^2 t^2/2}$, so

$M_Y^{(2)}(0) = a^2 + b^2$. Then $\text{Var}(Y) = (a^2 + b^2) - a^2 = b^2$

3.12.13 The moment generating function of Y is that of a normal variable with mean $\mu = -1$ and variance $\sigma^2 = 8$. Then $E(Y^2) = \text{Var}(Y) + \mu^2 = 8 + 1 = 9$.

3.12.15 $M_Y(t) = \int_a^b e^{ty} \frac{1}{b-a} dy = \frac{1}{(b-a)t} e^{ty} \Big|_a^b = \frac{1}{(b-a)t}(e^{tb} - e^{at})$ for $t \neq 0$

$$M_Y^{(1)}(t) = \frac{1}{(b-a)}\left[\frac{be^{tb} - ae^{at}}{t} - \frac{e^{tb} - e^{at}}{t^2}\right]$$

$$E(Y) = \lim_{t \to 0} M_Y^{(1)}(t) = \frac{1}{(b-a)} \lim_{t \to 0}\left[\frac{be^{tb} - ae^{at}}{t} - \frac{e^{tb} - e^{at}}{t^2}\right].$$ Applying L'Hospital's rule gives

$$E(Y) = \frac{1}{(b-a)}\left[(b^2 - a^2) - \frac{b^2 - a^2}{2}\right] = \frac{(a+b)}{2}$$

3.12.17 Let $Y = \frac{1}{\lambda}V$, where $f_V(y) = ye^{-y}$, $y \geq 0$. Question 3.12.8 establishes that $M_V(t) = \frac{1}{(1-t)^2}$. By Theorem 3.12.3(a), $M_Y(t) = M_V(t/\lambda) = 1(1-t/\lambda)^2$

3.12.19 (a) Let X and Y be two Poisson variables with parameters λ and μ respectively.
Then $M_X(t) = e^{-\lambda + \lambda e^t}$ and $M_Y(t) = e^{-\mu + \mu e^t}$. $M_{X+Y}(t) = M_X(t)M_Y(t)$
$= e^{-\lambda + \lambda e^t} e^{-\mu + \mu e^t} = e^{-(\lambda + \mu) + (\lambda + \mu)e^t}$. This last expression is that of a Poisson variable with parameter $\lambda + \mu$, which is then the distribution of $X + Y$.

(b) Let X and Y be two exponential variables, with parameters λ and μ respectively.
Then $M_X(t) = \frac{\lambda}{(\lambda - t)}$ and $M_Y(t) = \frac{\mu}{(\mu - t)}$.

$$M_{X+Y}(t) = M_X(t)M_Y(t) = \frac{\lambda}{(\lambda - t)}\frac{\mu}{(\mu - t)}.$$

This last expression is not that of an exponential variable, and the distribution of $X + Y$ is not exponential.

(c) Let X and Y be two normal variables, with parameters μ_1, σ_1^2 and μ_2, σ_2^2 respectively.
Then $M_X(t) = e^{\mu_1 t + \sigma_1^2 t^2 / 2}$ and $M_Y(t) = e^{\mu_2 t + \sigma_2^2 t^2 / 2}$.

$M_{X+Y}(t) = M_X(t)M_Y(t) = e^{\mu_1 t + \sigma_1^2 t^2 / 2} e^{\mu_2 t + \sigma_2^2 t^2 / 2} = e^{(\mu_1 + \mu_2)t + (\sigma_1^2 + \sigma_2^2)t^2 / 2}$.

This last expression is that of a normal variable with parameters $\mu_1 + \mu_2$ and σ_1^2 and σ_2^2, which is then the distribution of $X + Y$.

3.12.21 Let $S = \sum_{i=1}^n Y_i$. Then $M_S(t) = \prod_{i=1}^n M_{Y_i}(t) = \left(e^{\mu t + \sigma^2 t^2 / 2}\right)^n = e^{n\mu t + n\sigma^2 t^2 / 2}$. $M_{\bar{Y}}(t) = M_{S/n}(t) =$

$M_S(t/n) = e^{\mu t + (\sigma^2 / n)t^2 / 2}$. Thus $\bar{Y}$ is normal with mean μ and variance σ^2/n.

3.12.23 (a) $M_W(t) = M_{3X}(t) = M_X(3t) = e^{-\lambda + \lambda e^{3t}}$. This last term is not the moment-generating function of a Poisson random variable, so W is not Poisson.

(b) $M_W(t) = M_{3X+1}(t) = e^t M_X(3t) = e^t e^{-\lambda + \lambda e^{3t}}$. This last term is not the moment-generating function of a Poisson random variable, so W is not Poisson.

Chapter 4

Section 4.2

4.2.1 $p = P(\text{word is misspelled}) = \dfrac{1}{3250}$; $n = 6000$. Let x = number of words misspelled. Using

the exact binomial analysis, $P(X = 0) = \dbinom{6000}{0}\left(\dfrac{1}{3250}\right)^0 \left(\dfrac{3249}{3250}\right)^{6000} = 0.158$. For the Poisson

approximation, $\lambda = 6000\left(\dfrac{1}{3250}\right) = 1.846$, so $P(X = 0) \doteq \dfrac{e^{-1.846}(1.846)^0}{0!} = 0.158$. The

agreement is not surprising because n is so large and p is so small (recall Example 4.2.1).

4.2.3   Let X = number born on Poisson's birthday. Since $n = 500$, $p = \dfrac{1}{365}$, and $\lambda = 500 \cdot \dfrac{1}{365} =$

1.370, $P(X \le 1) = P(X = 0) + P(X = 1) \doteq \dfrac{e^{-1.370}(1.370)^0}{0!} + \dfrac{e^{-1.370}(1.370)^1}{1!} = 0.602$.

4.2.5   Let X = number of items requiring a price check. If $p = P(\text{item requires price check}) = 0.01$

and $n = 10$, a binomial analysis gives $P(X \ge 1) = 1 - P(X = 0) = 1 - \dbinom{10}{0}(0.01)^0(0.99)^{10} =$

0.10. Using the Poisson approximation, $\lambda = 10(0.01) = 0.1$ and $P(X \ge 1) = 1 - P(X = 0) \doteq$

$1 - \dfrac{e^{-0.1}(0.1)^0}{0!} = 0.10$. The exact model that applies here is the hypergeometric, rather than

the binomial, because p is a function of the previous items purchased. However, the variation
in p will be essentially zero for the 10 items purchased, so the binomial and hypergeometric
models in this case will be effectively the same.

4.2.7   Let X = number of pieces of luggage lost. Given that $n = 120$, $p = \dfrac{1}{200}$,

(so $\lambda = 120 \cdot \dfrac{1}{200} = 0.6$), $= P(X \ge 2) = 1 - P(X \le 1) = 1 - \displaystyle\sum_{k=0}^{1}\dfrac{e^{-0.6}(0.6)^k}{k!} = 0.122$.

4.2.9   Let X = number of solar systems with intelligent life and let $p = P(\text{solar system is inhabited})$.

For $n = 100{,}000{,}000{,}000$, $P(X \ge 1) = 1 - P(X = 0) = 1 - \dbinom{1{,}000{,}000{,}000}{0} p^0 \cdot (1 - p)^{100{,}000{,}000{,}000}$.

Solving $1 - (1 - p)^{1{,}000{,}000{,}000} = 0.50$ gives $p = 6.9 \times 10^{-12}$. Alternatively, it must be true that

$1 - \dfrac{e^{-\lambda}\lambda^0}{0!} = 0.50$, which implies that $\lambda = -\ln(0.50) = 0.69$. But $0.69 = np = 1 \times 10^{11} \cdot p$,

so $p = 6.9 \times 10^{-12}$.

4.2.11 The observed number of major changes $= 0.44$ $(= \bar{x} = \frac{1}{356}[237(0) + 90(1) + 22(2) + 7(3)])$,

so the presumed Poisson model is $p_X(k) = \dfrac{e^{-0.44}(0.44)^k}{k!}$, $k = 0, 1, \dots$ Judging from the

agreement evident in the accompanying table between the set of observed proportions and the values for $p_X(k)$, the hypothesis that X is a Poisson random variable is entirely credible.

No. of changes, k	Frequency	Proportion	$p_X(k)$
0	237	0.666	0.6440
1	90	0.253	0.2834
2	22	0.062	0.0623
3+	7	0.020	0.0102
	356	1.000	1.0000

4.2.13 The average of the data is $\dfrac{1}{113}[82(0) + 25(1) + 4(2) + 0(3) + 2(4)] \doteq 0.363$. Then use the model

$e^{-0.363}\dfrac{0.363^k}{k!}$. Usual statistical practice suggests collapsing the low frequency categories, in this case, $k = 2, 3, 4$. The result is the following table.

No. of countires, k	Frequency	$p_X(k)$	Expected frequency
0	82	0.696	78.6
1	25	0.252	28.5
2+	6	0.052	5.9

The level of agreement between the observed and expected frequencies suggests that the Poisson is a good model for these data.

4.2.15 If the mites exhibit any sort of "contagion" effect, the independence assumption implicit in

the Poisson model will be violated. Here, $\bar{x} = \dfrac{1}{100}[55(0) + 20(1) + \dots + 1(7)] = 0.81$, but

$p_X(k) = e^{-0.81}(0.81)^k/k!$, $k = 0, 1, \dots$ does not adequately approximate the infestation distribution.

No. of infestations, k	Frequency	Proportion	$p_X(k)$
0	55	0.55	0.4449
1	20	0.20	0.3603
2	21	0.21	0.1459
3	1	0.01	0.0394
4	1	0.01	0.0080
5	1	0.01	0.0013
6	0	0	0.0002
7+	1	0.01	0.0000
		1.00	1.0000

4.2.17 Let X = number of transmission errors made in next half-minute. Since $E(X) = \lambda = 4.5$,

$$P(X > 2) = 1 - P(X \le 2) = 1 - \sum_{k=0}^{2} \frac{e^{-4.5}(4.5)^k}{k!} = 0.826.$$

4.2.19 Let X = number of flaws in 40 sq. ft. Then $E(X) = 4$ and $P(X \ge 3) = 1 - P(X \le 2) =$

$$1 - \sum_{k=0}^{2} \frac{e^{-4} 4^k}{k!} = 0.762.$$

4.2.21 (a) Let X = number of accidents in next five days. Then $E(X) = 0.5$ and
$P(X = 2) = e^{-0.5}(0.5)^2/2! = 0.076.$

(b) No. P(4 accidents occur during next two weeks) $=$
$P(X = 4) \cdot P(X = 0) + P(X = 3) \cdot P(X = 1) + P(X = 2) \cdot P(X = 2) + P(X = 1) \cdot P(X = 3) + P(X = 0) \cdot P(X = 4).$

4.2.23 $P(X \text{ is even}) = \sum_{k=0}^{\infty} \frac{e^{-\lambda}\lambda^{2k}}{(2k)!} = e^{-\lambda}\left\{1 + \frac{\lambda^2}{2!} + \frac{\lambda^4}{4!} + \frac{\lambda^6}{6!} + \cdots\right\} = e^{-\lambda} \cdot \cosh \lambda =$

$$e^{-\lambda}\left(\frac{e^{\lambda} + e^{-\lambda}}{2}\right) = \frac{1}{2}(1 + e^{-2\lambda}).$$

4.2.25 From Definition 3.11.1 and Theorem 3.7.1, $P(X_2 = k) = \sum_{x_1 = k}^{\infty} \binom{x_1}{k} p^k (1-p)^{x_1 - k} \cdot \frac{e^{-\lambda}\lambda^{x_1}}{x_1!}.$

Let $y = x_1 - k$. Then $P(X_2 = k) = \sum_{y=0}^{\infty} \binom{y+k}{k} p^k (1-p)^y \cdot \frac{e^{-\lambda}\lambda^{y+k}}{(y+k)!} = \frac{e^{-\lambda}(\lambda p)^k}{k!} \cdot \sum_{y=0}^{\infty} \frac{[\lambda(1-p)]^y}{y!} =$

$$\frac{e^{-\lambda}(\lambda p)^k}{k!} \cdot e^{\lambda(1-p)} = \frac{e^{-\lambda p}(\lambda p)^k}{k!}.$$

4.2.27 Let X = number of deaths in a week. Based on the daily death rate, $E(X) = \lambda = 0.7$. Let Y = interval (in weeks) between consecutive deaths. Then $P(Y > 1) = \int_1^{\infty} 0.7 e^{-0.7y} dy = -e^{-u} \Big|_{0.7}^{\infty} = 0.50.$

4.2.29 Let X = number of bulbs burning out in 1 hr. Then $E(X) = \lambda = 0.11$. Let Y = number of hours a bulb remains lit. Then $P(Y < 75) = \int_0^{75} 0.011 e^{-0.011y} dy = -e^{-u} \Big|_0^{0.825} = 0.56.$ (where $u = 0.011y$). Since $n = 50$ bulbs are initially online, the expected number that will fail to last at least 75 hours is $50 \cdot P(Y < 75)$, or 28.

Section 4.3

4.3.1   a)   0.5782   b)   0.8264   c)   0.9306   d)   0.0000

4.3.3   a)   Both are the same because of the symmetry of $f_Z(z)$.

b)   Since $f_Z(z)$ is decreasing for all $z > 0$, $\int_{a-\frac{1}{2}}^{a+\frac{1}{2}} \frac{1}{\sqrt{2\pi}} e^{-z^2/2} dz$ is larger than

$\int_{a}^{a+1} \frac{1}{\sqrt{2\pi}} e^{-z^2/2} dz$.

4.3.5   a) -0.44 b)   0.76   c)   0.41   d)   1.28   e)   0.95

4.3.7   Let X = number of decals purchased in November. Then X is binomial with $n = 74,806$ and $p = 1/12$.
$P(50X < 306,000) = P(X < 6120) = P(X \le 6119)$. Using the DeMoivre-Laplace approximation with continuity correction gives

$$P(X \le 6119) \doteq P\left(Z \le \frac{6119.5 - 74,806(1/12)}{\sqrt{74,806(1/12)(11/12)}} \right) = P(Z \le -1.51) = 0.0655$$

4.3.9   Let X = number of voters challenger receives. Given that $n = 400$ and $p = P$(voter favors challenger) $= 0.45$, $np = 180$ and $np(1 - p) = 99$.

a) $P(\text{tie}) = P(X = 200) = P(199.5 \le X \le 200.5) =$
$$P\left(\frac{199.5 - 180}{\sqrt{99}} \le \frac{X - 180}{\sqrt{99}} \le \frac{200.5 - 180}{\sqrt{99}} \right) \doteq P(1.96 \le Z \le 2.06) = 0.0053.$$

b) $P(\text{challenger wins}) = P(X > 200) = P(X \ge 200.5) =$
$$P\left(\frac{X - 180}{\sqrt{99}} \ge \frac{200.5 - 180}{\sqrt{99}} \right) \doteq P(Z \ge 2.06) = 0.0197.$$

4.3.11   Let $p = P$(person dies by chance in the three months following birthmonth) $= \dfrac{1}{4}$. Given that $n = 747$, $np = 186.75$, and $np(1 - p) = 140.06$, $P(X \ge 344) = P(X \ge 343.5) =$
$$P\left(\frac{X - 186.75}{\sqrt{140.06}} \ge \frac{343.5 - 186.75}{\sqrt{140.06}} \right) = P(Z \ge 13.25) = 0.0000.$$ The fact that the latter
probability is so small strongly discredits the hypothesis that people die randomly with respect to their birthdays.

4.3.13   No, the normal approximation is inappropriate because the values of n (= 10) and p (= 0.7) fail to satisfy the condition $n > 9\dfrac{p}{1-p} = 9\dfrac{0.7}{0.3} = 21$.

4.3.15 $P(|X - E(X)| \le 5) = P(-5 \le X - 100 \le 5) = P\left(\dfrac{-5.5}{\sqrt{50}} \le \dfrac{X - 100}{\sqrt{50}} \le \dfrac{5.5}{\sqrt{50}}\right) \doteq$

$P(-0.78 \le Z \le 0.78) = 0.5646.$

For binomial data, the central limit theorem and DeMoivre-Laplace approximations differ only if the continuity correction is used in the DeMoivre-Laplace approximation.

4.3.17 For the given X, $E(X) = 5(18/38) + (-5)(20/38) = -10/38 = -0.263.$
$\text{Var}(X) = 25(18/38) + (25)(20/38) - (-10/38)^2 = 24.931$, $\sigma = 4.993.$

Then $P(X_1 + X_2 + \ldots + X_{100} > -50)$

$= P\left(\dfrac{X_1 + X_2 + \ldots + X_{100} - 100(-0.263)}{\sqrt{100}(4.993)} > \dfrac{-50 - 100(-0.263)}{10(4.993)}\right) \doteq P(Z > -1.53)$

$= 1 - 0.0630 = 0.9370$

4.3.19 Let X = number of chips ordered next week. Given that $\lambda = E(X) = 50$, P(company is unable to fill orders) $= P(X \ge 61) = P(X \ge 60.5) = P\left(\dfrac{X - 50}{\sqrt{50}} \ge \dfrac{60.5 - 50}{\sqrt{50}}\right) \; P(Z \ge 1.48) = 0.0694.$

4.3.21 No, only 84% of drivers are likely to get at least 25,000 miles on the tires. If X denotes the mileage obtained on a set of Econo-Tires, $P(X \ge 25,000) =$

$P\left(\dfrac{X - 30,000}{5000} \ge \dfrac{25,000 - 30,000}{5000}\right) = P(Z \ge -1.00) = 0.8413.$

4.3.23 Let Y = donations collected tomorrow. Given that $\mu = \$20,000$ and $\sigma = \$5,000$,

$P(Y > \$30,000) = P\left(\dfrac{Y - \$20,000}{\$5,000} > \dfrac{\$30,000 - \$20,000}{\$5,000}\right) = P(Z > 2.00) = 0.0228.$

4.3.25 a) Let Y_1 and Y_2 denote the scores made by a random nondelinquent and delinquent, respectively. Then $E(Y_1) = 60$ and $\text{Var}(Y_1) = 10^2$; also, $E(Y_2) = 80$ and $\text{Var}(Y_2) = 5^2$. Since 75 is the cutoff between teenagers classified as delinquents or nondelinquents,

P(nondelinquent is misclassified as delinquent) $= P(Y_1 > 75) = P\left(Z > \dfrac{75 - 60}{10}\right) =$

0.0668. Similarly, P(delinquent is misclassified as nondelinquent) $= P(Y_2 < 75) =$

$P\left(Z < \dfrac{75 - 80}{5}\right) = 0.1587.$

4.3.27 Let Y = freshman's verbal SAT score. Given that $\mu = 565$ and $\sigma = 75$, $P(Y > 660) =$

$P\left(\dfrac{Y - 565}{75} > \dfrac{660 - 565}{75}\right) = P(Z > 1.27) = 0.1020.$ It follows that the expected <u>number</u> doing better is 4250(0.1020), or 434.

4.3.29 If $P(20 \leq Y \leq 60) = 0.50$, then $P\left(\dfrac{20-40}{\sigma} \leq \dfrac{Y-40}{\sigma} \leq \dfrac{60-40}{\sigma}\right) = 0.50 =$

$P\left(\dfrac{-20}{\sigma} \leq Z \leq \dfrac{20}{\sigma}\right)$. But $P(-0.67 \leq Z \leq 0.67) = 0.4972 \doteq 0.50$, which implies that $0.67 =$

$\dfrac{20}{\sigma}$. The desired value for σ, then, is $\dfrac{20}{0.67}$, or 29.85.

4.3.31 Let Y = analyzer reading for driver whose true blood alcohol concentration is 0.11. Then

$P(\text{analyzer mistakenly shows driver to be sober}) = P(Y < 0.10) = P\left(\dfrac{Y-0.11}{0.004} < \dfrac{0.10-0.11}{0.004}\right)$

$= P(Z < -2.50) = 0.0062$. The "0.095%) driver should ask to take the test twice. The "0.11" driver has a greater chance of not being charged by taking the test only once. As, n the number of times the test taken, increases, the precision of the average reading increases. It is to the sober driver's advantage to have a reading as precise as possible; the opposite is true for the drunk driver.

4.3.33 By the first corollary to Theorem 4.3.3, $P(\bar{Y} > 103) = P\left(\dfrac{\bar{Y}-100}{16/\sqrt{9}} > \dfrac{103-100}{16/\sqrt{9}}\right) = P(Z > 0.56)$

$= 0.2877$. For any arbitrary Y_i, $P(Y_i > 103) = P\left(\dfrac{Y_i-100}{16} > \dfrac{103-100}{16}\right) = P(Z > 0.19) =$

0.4247.
Let X = number of Y_i's that exceed 103. Since X is a binomial random variable with $n = 9$

and $p = P(Y_i > 103) = 0.4247$, $P(X = 3) = \dbinom{9}{3}(0.4247)^3(0.5753)^6 = 0.23$.

4.3.35 Let Y_i = resistance of ith resistor, $i = 1, 2, 3$, and let $Y = Y_1 + Y_2 + Y_3$ = circuit resistance. By the first corollary to Theorem 4.3.3, $E(Y) = 6 + 6 + 6 = 18$ and $Var(Y) = (0.3)^2 + (0.3)^2 +$

$(0.3)^2 = 0.27$. Therefore, $P(Y > 19) = P\left(\dfrac{Y-18}{\sqrt{0.27}} > \dfrac{19-18}{\sqrt{0.27}}\right) = P(Z > 1.92) = 0.0274$. Suppose

$P(Y > 19) = 0.005 = P\left(Z > \dfrac{19-18}{\sqrt{3\sigma^2}}\right)$. From Appendix Table A.1, $P(Z > 2.58) \doteq 0.005$, so

$2.58 = \dfrac{19-18}{\sqrt{3\sigma^2}}$, which implies that the minimum "precision" of the manufacturing process

would have to be $\sigma = 0.22$ ohms.

4.3.37 $M_{\bar{Y}}(t) = M_{Y_1 + \cdots Y_n}\left(\dfrac{t}{n}\right) = \prod_{i=1}^{n} M_{Y_i}\left(\dfrac{t}{n}\right) = \prod_{i=1}^{n} e^{\mu t/n + \sigma^2 t^2/2n^2} = e^{\mu t + \sigma^2 t^2/2n}$, but the latter is the

moment-generating function for a normal random variable whose mean is μ and whose

variance is σ^2/n. Similarly, if $Y = a_1 Y_1 + \ldots + a_n Y_n$, $M_Y(t) = \prod_{i=1}^{n} M_{a_i Y_i}(t) = \prod_{i=1}^{n} M_{Y_i}(a_i t) =$

$\prod_{i=1}^{n} e^{\mu_i a_i t + \sigma_i^2 a_i^2 t^2/2} = e^{\sum_{i=1}^{n} a_i \mu_i t + \sum_{i=1}^{n} a_i^2 \sigma_i^2 t^2/2}$. By inspection, Y has the moment-generating

function of a normal random variable for which $E(Y) = \sum_{i=1}^{n} a_i \mu_i$ and $\mathrm{Var}(Y) = \sum_{i=1}^{n} a_i^2 \sigma_i^2$.

Section 4.4

4.4.1 Let $p = P(\text{return is audited in a given year}) = 0.30$ and let $X = $ year of first audit. Then
$P(\text{Jody escapes detection for at least 3 years}) = P(X \geq 4) = 1 - P(X \leq 3) =$

$1 - \sum_{k=1}^{3} (0.70)^{k-1}(0.30) = 0.343$.

4.4.3 No, the expected frequencies ($= 5 \cdot p_X(k)$) differ considerably from the observed frequencies, especially for small values of k. The observed number of 1's, for example, is 4, while the expected number is 12.5.

k	Obs. Freq.	$p_X(k) = \left(\dfrac{3}{4}\right)^{k-1}\left(\dfrac{1}{4}\right)$	$50 \cdot p_X(k) = $ Exp. freq.
1	4	0.2500	12.5
2	13	0.1875	9.4
3	10	0.1406	7.0
4	7	0.1055	5.3
5	5	0.0791	4.0
6	4	0.0593	3.0
7	3	0.0445	2.2
8	3	0.0334	1.7
9+	1	0.1001	5.0
	50	1.0000	50.0

4.4.5 $F_X(t) = P(X \leq t) = p \sum_{s=0}^{[t]} (1-p)^s$. But $\sum_{s=0}^{[t]} (1-p)^s = \dfrac{1-(1-p)^{[t]}}{1-(1-p)} = \dfrac{1-(1-p)^{[t]}}{p}$, and the

result follows.

4.4.7 $P(n \le Y \le n+1) = \int_n^{n+1} \lambda e^{-\lambda y}dy = (1 - e^{-\lambda y})\Big|_n^{n+1} = e^{-\lambda n} - e^{-\lambda(n+1)} = e^{-\lambda n}(1 - e^{-\lambda})$

Setting $p = 1 - e^{-\lambda}$ gives $P(n \le Y \le n+1) = (1-p)^n p$.

4.4.9 $M_X(t) = pe^t[1 - (1-p)e^t]^{-1}$, so $M_X^{(1)}(t) = pe^t(-1)[1 - (1-p)e^t]^{-2} \cdot (-(1-p)e^t) +$

$[1 - (1-p)e^t]^{-1}pe^t$. Setting $t = 0$ gives $M_X^{(1)}(0) = E(X) = \dfrac{1}{p}$. Similarly, $M_X^{(2)}(t) =$

$p(1-p)e^{2t}(-2)[1 - (1-p)e^t]^{-3} \cdot (-(1-p)e^t) + [1 - (1-p)e^t]^{-2}p(1-p)e^{2t} \cdot 2 +$

$[1 - (1-p)e^t]^{-1}pe^t + pe^t(-1)[1 - (1-p)e^t]^{-2} \cdot (-(1-p)e^t)$ and $M_X^{(2)}(0) = E(X^2) = \dfrac{2-p}{p^2}$.

Therefore, $\mathrm{Var}(X) = E(X^2) - [E(X)]^2 = \dfrac{2-p}{p^2} - \left(\dfrac{1}{p}\right)^2 = \dfrac{1-p}{p^2}$

4.4.11 Let $M_X^{\bullet}(t) = E(t^X) = \displaystyle\sum_{k=1}^{\infty} t^k \cdot (1-p)^{k-1}p = \dfrac{p}{1-p}\sum_{k=1}^{\infty}[t(1-p)]^k = \dfrac{p}{1-p}\sum_{k=0}^{\infty}[t(1-p)]^k - \dfrac{p}{1-p} =$

$\dfrac{p}{1-p}\left[\dfrac{1}{1-t(1-p)}\right] - \dfrac{p}{1-p} = \dfrac{pt}{1-t(1-p)} = $ factorial moment-generating function for X. Then

$M_X^{\bullet(1)}(t) = pt(-1)[1 - t(1-p)]^{-2}(-(1-p)) + [1 - t(1-p)]^{-1}p = \dfrac{p}{[1-t(1-p)]^2}$. When $t = 1$,

$M_X^{\bullet(1)}(1) = E(X) = \dfrac{1}{p}$. Also, $M_X^{\bullet(2)}(t) = \dfrac{2p(1-p)}{[1-t(1-p)]^3}$ and $M_X^{\bullet(2)}(1) = \dfrac{2-2p}{p^2} =$

$E[X(X-1)] = E(X^2) - E(X)$. Therefore, $\mathrm{Var}(X) = E(X^2) - [E(X)]^2 = \dfrac{2-2p}{p^2} + \dfrac{1}{p} - \left(\dfrac{1}{p}\right)^2$

$= \dfrac{1-p}{p^2}$.

Section 4.5

4.5.1 Let X = number of houses needed to achieve fifth invitation. If $p = P$(saleswoman receives invitation at a given house) $= 0.30$, $p_X(k) = \dbinom{k-1}{4}(0.30)^4(0.70)^{k-1-4}(0.30)$, $k = 5, 6, \ldots$ and

$P(X < 8) = P(5 \leq X \leq 7) = \displaystyle\sum_{k=5}^{7}\dbinom{k-1}{4}(0.30)^5(0.70)^{k-5} = 0.029.$

4.5.3 Darryl might have actually done his homework, but there is reason to suspect that he did not. Let the random variable X denote the toss where a head appears for the second time. Then
$$p_X(k) = \binom{k-1}{1}\left(\frac{1}{2}\right)^2\left(\frac{1}{2}\right)^{k-2}, \ k = 2, 3, \ldots,$$ but that particular model fits the data almost perfectly, as the table shows. Agreement this good is often an indication that the data have been fabricated.

k	$p_X(k)$	Obs. freq.	Exp. freq.
2	1/4	24	25
3	2/8	26	25
4	3/16	19	19
5	4/32	13	12
6	5/64	8	8
7	6/128	5	5
8	7/256	3	3
9	8/512	1	2
10	9/1024	1	1

4.5.5 $E(X) = \sum_{k=r}^{\infty} k\binom{k-1}{r-1}p^r(1-p)^{k-r} = \frac{r}{p}\sum_{k=r}^{\infty}\binom{k}{r}p^{r+1}(1-p)^{k-r} = \frac{r}{p}.$

4.5.7 Here $X = Y - r$, where Y has the negative binomial distribution as described in Theorem 4.5.1. Using the properties (1), (2), and (3) given by the theorem, we can write $E(X) = E(Y - r) = E(Y)$
$- E(r) = \frac{r}{p} - r = \frac{r(1-p)}{p}$ and $\text{Var}(X) = \text{Var}(Y - r) = \text{Var}(Y) + \text{Var}(r) = \frac{r(1-p)}{p^2} + 0 = \frac{r(1-p)}{p^2}.$

Also, $M_X(t) = M_{Y-r}(t) = e^{-rt}M_Y(t) = e^{-rt}\left[\frac{pe^t}{1-(1-p)e^t}\right]^r = \left[\frac{p}{1-(1-p)e^t}\right]^r.$

4.5.9 $M_X^{(1)}(t) = r\left[\frac{pe^t}{1-(1-p)e^t}\right]^{r-1}[pe^t[1-(1-p)e^t]^{-2}(1-p)e^t + [1-(1-p)e^t]^{-1}pe^t].$ When $t = 0$,

$M_X^{(1)}(0) = E(X) = r\left[\frac{p(1-p)}{p^2} + \frac{p}{p}\right] = \frac{r}{p}.$

Section 4.6

4.6.1 Let Y_i = lifetime of ith gauge, $i = 1, 2, 3$. By assumption, $f_{Y_i}(y) = 0.001e^{-0.001y}, y > 0$. Define the random variable $Y = Y_1 + Y_2 + Y_3$ to be the lifetime of the system. By Theorem 4.6.1, $f_Y(y) = \frac{(0.001)^3}{2}y^2e^{-0.001y}, y > 0$.

4.6.3 The time until the 24th breakdown is a gamma random variable with parameters $r = 24$ and $\lambda = 3$. The mean of this random variable is $r/\lambda = 24/3 = 8$ months.

4.6.5 $f_{\lambda Y}(y) = \dfrac{1}{\lambda} f_Y(y/\lambda) = \dfrac{1}{\lambda} \dfrac{\lambda^r}{\Gamma(r)}\left(\dfrac{y}{\lambda}\right)^{r-1} e^{-\lambda(y/\lambda)} = \dfrac{1}{\Gamma(r)} y^{r-1} e^{-y}$

4.6.7 $\Gamma\left(\dfrac{7}{2}\right) = \dfrac{5}{2}\Gamma\left(\dfrac{5}{2}\right) = \dfrac{5}{2}\dfrac{3}{2}\Gamma\left(\dfrac{3}{2}\right) = \dfrac{5}{2}\dfrac{3}{2}\dfrac{1}{2}\Gamma\left(\dfrac{1}{2}\right) = \dfrac{15}{8}\Gamma\left(\dfrac{1}{2}\right)$ by Theorem 4.6.2(2)

Further, $\Gamma\left(\dfrac{1}{2}\right) = \sqrt{\pi}$ by Question 4.6.6.

4.6.9   Write the gamma moment-generating function in the form $M_Y(t) = (1 - t/\lambda)^{-r}$. Then $M_Y^{(1)}(t) = -r(1 - t/\lambda)^{-r-1}(-1/\lambda) = (r/\lambda)(1 - t/\lambda)^{-r-1}$ and $M_Y^{(2)}(t) = (r/\lambda)(-r - 1)(1 - t/\lambda)^{-r-2} \cdot (-1/\lambda) = (r/\lambda^2)(r + 1)(1 - t/\lambda)^{-r-2}$. Therefore, $E(Y) = M_Y^{(1)}(0) = \dfrac{r}{\lambda}$ and $\mathrm{Var}(Y) = M_Y^{(2)}(0) - \left[M_Y^{(1)}(0)\right]^2 = \dfrac{r(r+1)}{\lambda^2} - \dfrac{r^2}{\lambda^2} = \dfrac{r}{\lambda^2}.$

Chapter 5

Section 5.2

5.2.1 $L(\theta) = \prod_{i=1}^{8} \theta^{k_i}(1-\theta)^{1-k_i} = \theta^{\sum_{i=1}^{8} k_i}(1-\theta)^{8-\sum_{i=1}^{8} k_i} = \theta^5(1-\theta)^3$

$\dfrac{dL(\theta)}{d\theta} = \theta^5 3(1-\theta)^2(-1) + 5\theta^4(1-\theta)^3 = \theta^4(1-\theta)^2(-8\theta+5) \cdot \dfrac{dL(\theta)}{d\theta} = 0$ implies $\theta_e = 5/8$

5.2.3 $L(\theta) = \prod_{i=1}^{4} \lambda e^{-\lambda y_i} = \lambda^4 e^{-\lambda \sum_{i=1}^{4} y_i} = \lambda^4 e^{-32.8\lambda}.$

$\dfrac{dL(\lambda)}{d\lambda} = \lambda^4(-32.8)e^{-32.8\lambda} + 4\lambda^3 e^{-32.8\lambda} = \lambda^3 e^{-32.8\lambda}(4 - 32.8\lambda)$

$\dfrac{dL(\lambda)}{d\lambda} = 0$ implies $\lambda_e = 4/32.8 = 0.122$

5.2.5 $L(\theta) = \prod_{i=1}^{3} \dfrac{y_i^3 e^{-y_i/\theta}}{6\theta^4} = \dfrac{\left(\prod_{i=1}^{3} y_i^3\right) e^{-\sum_{i=1}^{3} y_i/\theta}}{216\theta^{12}}.$

$\ln L(\theta) = \ln \prod_{i=1}^{3} y_i^3 - \dfrac{1}{\theta}\sum_{i=1}^{3} y_i - \ln 216 - 12 \ln \theta$

$\dfrac{d\ln L(\theta)}{d\theta} = \dfrac{1}{\theta^2}\sum_{i=1}^{3} y_i - \dfrac{12}{\theta} = \dfrac{\sum_{i=1}^{3} y_i - 12\theta}{\theta^2}$

or $\dfrac{d\ln L(\theta)}{d\theta} = 0$ implies $\dfrac{\sum_{i=1}^{3} y_i - 12\theta}{\theta^2} = \dfrac{8.8 - 12\theta}{\theta^2} = 0$

or $\theta_e = 0.733$

5.2.7 $L(\theta) = \prod_{i=1}^{5} \theta y_i^{\theta-1} = \theta^5 \left(\prod_{i=1}^{5} y_i\right)^{\theta-1}.$

$\ln L(\theta) = 5 \ln \theta + (\theta-1)\sum_{i=1}^{5} \ln y_i$

$\dfrac{d\ln L(\theta)}{d\theta} = \dfrac{5}{\theta} + \sum_{i=1}^{5} \ln y_i = \dfrac{5 + \theta\sum_{i=1}^{5} \ln y_i}{\theta}$

$\dfrac{d\ln L(\theta)}{d\theta} = 0$ implies $\dfrac{5 - 0.625\theta}{\theta} = 0$ or $\theta_e = 8.00$

5.2.9 a) $L(\theta) = \left(\dfrac{1}{\theta}\right)^n$, if $0 \le y_1, y_2, \ldots, y_n \le \theta$, and 0 otherwise. Thus $\theta_e = y_{max}$, which for these data is 14.2.

b) $L(\theta) = \left(\dfrac{1}{\theta_2 - \theta_1}\right)^n$, if $\theta_1 \le y_1, y_2, \ldots, y_n \le \theta_2$, and 0 otherwise. Thus $\theta_{1e} = y_{min}$ and $\theta_{2e} = y_{max}$. For these data, $\theta_{1e} = 1.8$, $\theta_{2e} = 14.2$.

5.2.11 $L(\theta) = \displaystyle\prod_{i=1}^{n} 2y_i \theta^2 = 2^n \left(\prod_{i=1}^{n} y_i\right)\theta^{2n}$, if $0 \le y_1, y_2, \ldots, y_n \le 1/\theta$, and 0 otherwise. To maximize $L(\theta)$ maximize θ. Since each $y_i \le 1/\theta$, then $\theta \le 1/y_i$ for $1 \le i \le n$. Thus, the maximum value for θ under these constraints is the minimum of the $1/y_i$, or $\theta_e = 1/y_{max}$.

5.2.13 (a) $L(\alpha, \beta) = \displaystyle\prod_{i=1}^{n} \alpha\beta y_i^{\beta-1} e^{-\alpha y_i^\beta} = \alpha^n \beta^n \left(\prod_{i=1}^{n} y_i\right)^{\beta-1} e^{-\alpha \sum_{i=1}^{n} y_i^\beta}$

$\ln L(\alpha, \beta) = n \ln \alpha + n \ln \beta + (\beta - 1) \ln\left(\displaystyle\prod_{i=1}^{n} y_i\right) - \alpha \sum_{i=1}^{n} y_i^\beta$

$\dfrac{\partial \ln L(\alpha, \beta)}{\partial \alpha} = \dfrac{n}{\alpha} - \displaystyle\sum_{i=1}^{n} y_i^\beta$

Setting $\dfrac{\partial \ln L(\alpha, \beta)}{\partial \alpha} = 0$ gives $\alpha_e = \dfrac{n}{\displaystyle\sum_{i=1}^{n} y_i^\beta}$

(b) The first equation is line two of part (a)
To obtain the other equation, consider

$\dfrac{\partial \ln L(\alpha, \beta)}{\partial \beta} = \dfrac{n}{\beta} + \ln \displaystyle\prod_{i=1}^{n} y_i - \alpha \sum_{i=1}^{n} \beta y_i^{\beta-1}$

Setting $\dfrac{\partial \ln L(\alpha, \beta)}{\partial \alpha} = 0$ provides the other equation.
Solving the two simultaneously would be done by approximation methods.

5.2.15 For Y uniform on $(0, \theta)$, $E(Y) = \theta/2$. Set $\theta/2 = \bar{y}$, so the method of moments estimate is $\theta_e = 2\bar{y}$. For the data given, $\theta_e = 2(50) = 100$. The maximum likelihood estimate is $y_{max} = 92$.

5.2.17 For Y Poisson, $E(Y) = \lambda$. Then $\lambda_e = \bar{y} = 13/6$. The maximum likelihood estimate is the same.

5.2.19 $E(Y) = \theta_1$ so $\theta_{1e} = \bar{y}$.

$$E(Y^2) = \int_{\theta_1-\theta_2}^{\theta_1+\theta_2} y^2 \frac{1}{2\theta_2} dy = \frac{1}{2\theta_2} \left[\frac{y^3}{3} \right]_{\theta_1-\theta_2}^{\theta_1+\theta_2} = \theta_1^2 + \frac{1}{3}\theta_2^2$$

Substitute $\theta_{1e} = \bar{y}$ into the equation $\theta_1^2 + \frac{1}{3}\theta_2^2 = \frac{1}{n}\sum_{i=1}^{n} y_i^2$ to obtain

$$\theta_{2e} = \sqrt{3\left(\frac{1}{n}\sum_{i=1}^{n} y_i^2 - \bar{y}^2 \right)}.$$

5.2.21 $E(X) = 0 \cdot \theta^0 (1-\theta)^{1-0} + 1 \cdot \theta^1 (1-\theta)^{1-1} = \theta.$
Then $\theta_e = \bar{y}$, which for the given data is 2/5.

5.2.23 From Theorem 4.5.1, $E(X) = r/p$. $\mathrm{Var}(X) = \dfrac{r(1-p)}{p^2}$.

Then $E(X^2) = \mathrm{Var}(X) + E(X)^2 = \dfrac{r(1-p)+r^2}{p^2}$

Set $\dfrac{r}{p} = \bar{x}$ to obtain $r = p\bar{x}$, and substitute into the equation

$\dfrac{r(1-p)+r^2}{p^2} = \dfrac{1}{n}\sum_{i=1}^{n} x_i^2$ to obtain an equation in p: $\dfrac{p\bar{x}(1-p)+(p\bar{x})^2}{p^2} = \dfrac{1}{n}\sum_{i=1}^{n} x_i^2$.

Equivalently, $p\bar{x} - p^2\bar{x} + p^2\bar{x}^2 - p^2\dfrac{1}{n}\sum_{i=1}^{n} x_i^2 = 0$. Solving for p gives $p_e = \dfrac{\bar{x}}{\bar{x} + \dfrac{1}{n}\sum_{i=1}^{n} x_i^2 - \bar{x}^2}$.

Then $r_e = \dfrac{\bar{x}^2}{\bar{x} + \dfrac{1}{n}\sum_{i=1}^{n} x_i^2 - \bar{x}^2}$.

Section 5.3

5.3.1 The confidence interval is $\left(\bar{y} - z_{\alpha/2}\dfrac{\sigma}{\sqrt{n}}, \bar{y} + z_{\alpha/2}\dfrac{\sigma}{\sqrt{n}} \right) =$

$\left(0.766 - 1.96\dfrac{0.09}{\sqrt{19}}, 0.766 + 1.96\dfrac{0.09}{\sqrt{19}} \right) = (0.726, 0.806).$
The value of 0.80 is believable.

5.3.3 The length of the confidence interval is

$2z_{\alpha/2}\dfrac{\sigma}{\sqrt{n}} = \dfrac{2(1.96)(14.3)}{\sqrt{n}} = \dfrac{56.056}{\sqrt{n}}$. For $\dfrac{56.056}{\sqrt{n}} \le 3.06$, $n \ge \left(\dfrac{56.056}{3.06} \right)^2 = 335.58$, so take $n = 336$.

5.3.5 The probability that the given interval will contain μ is $P(-0.96 < Z < 1.06) = 0.6869$. The probability of four or five such intervals is binomial with $n = 5$ and $p = 0.6869$, so the probability is $5(0.6869)^4(0.3131) + (0.6869)^5 = 0.501$.

5.3.7 The interval given is correctly *calculated*. However, the data do not appear to be normal, so claiming that it is a 95% confidence interval would not be correct.

5.3.9 $\left(\dfrac{192}{540} - 1.96\sqrt{\dfrac{(192/540)(1-192/540)}{540}}, \ \dfrac{192}{540} + 1.96\sqrt{\dfrac{(192/540)(1-192/540)}{540}} \right)$
$= (0.316, 0.396)$

5.3.11 Budweiser would use the sample proportion 0.54 alone as the estimate. Schlitz would construct the 95% confidence interval $(0.36, 0.56)$ to claim that values < 0.50 are believable.

5.3.13 $2.58\sqrt{\dfrac{p(1-p)}{n}} \le 2.58\sqrt{\dfrac{1}{4n}} \le 0.01$, so take $n \ge \dfrac{(2.58)^2}{4(0.01)^2} = 16{,}641$

5.3.15 Both intervals have confidence level approximately 50%.

5.3.17 $\dfrac{z_a}{2\sqrt{1013}} = \dfrac{3.1}{100}$ implies $z_a = 1.97$. The margin of error is correct at the 95% level. For the given data, estimates of the percentage as small as $0.30 - 0.031 = 0.269$ or as large as $0.30 + 0.031 = 0.331$ are believable.

5.3.19 If X is hypergeometric, then $\text{Var}(X/n) = \dfrac{p(1-p)}{n} \dfrac{N-n}{N-1}$.

As before $p(1-p) \le 1/4$. Thus, in Definition 5.3.1, substitute $d = \dfrac{1.96}{2\sqrt{n}}\sqrt{\dfrac{N-n}{N-1}}$.

5.3.21 If n is such that $0.06 = \dfrac{1.96}{2\sqrt{n}}$, then n is the smallest integer $\ge \dfrac{1.96^2}{4(0.06)^2} = 266.8$.
Take $n = 267$.

If n is such that $0.03 = \dfrac{1.96}{2\sqrt{n}}$, then n is the smallest integer $\ge \dfrac{1.96^2}{4(0.03)^2} = 1067.1$.
Take $n = 1068$.

5.3.23 Case 1: n is the smallest integer greater than
$\dfrac{z_{.02}^2}{4(0.05)^2} = \dfrac{2.05^2}{4(0.05)^2} = 420.25$, so take $n = 421$.
Case 2: n is the smallest integer greater than
$\dfrac{z_{.04}^2}{4(0.04)^2} = \dfrac{1.75^2}{4(0.04)^2} = 478.5$, so take $n = 479$.

5.3.25 Take n to be the smallest integer $\geq \dfrac{z_{10}^2}{4(0.02)^2} = \dfrac{1.28^2}{4(0.02)^2} = 1024.$

Section 5.4

5.4.1 $P(|\hat{\theta} - 3| > 1.0) = P(\hat{\theta} < 2) + P(\hat{\theta} > 4)$
$= P(\hat{\theta} = 1.5) + P(\hat{\theta} = 4.5) = P((1,2)) + P((4,5)) = 2/10$

5.4.3 $P(X < 250) = P\left(\dfrac{X - 500(0.52)}{\sqrt{500(0.52)(0.48)}} < \dfrac{250 - 500(0.52)}{\sqrt{500(0.52)(0.48)}}\right) = P(Z < -0.90) = 0.1841$

5.4.5 a) $E(\overline{X}) = E\left(\dfrac{1}{n}\sum_{i=1}^{n} X_i\right) = \dfrac{1}{n}\sum_{i=1}^{n} E(X_i) = \dfrac{1}{n}\sum_{i=1}^{n} \lambda = \lambda$

b) In general, the sample mean is an unbiased estimator of the mean μ.

5.4.7 First note that $E(Y) = \int_{\theta}^{\infty} y e^{-(y-\theta)} dy \int_{0}^{\infty}(u+0)e^u du = \int_{0}^{\infty} ue^u du + \theta\int_{0}^{\infty} e^u du = 1 + \theta$

Then $E(\overline{Y}) = E(Y) = 1 + \theta$, so $E(\overline{Y} - 1) = \theta$.

5.4.9 $E(Y) = 2\int_{0}^{1/\theta} y^2 \theta^2 dy = \dfrac{2}{3}\left(\dfrac{1}{\theta}\right).$

$E[c(Y_1 + 2Y_2)] = c[E(Y_1) + 2E(Y_2)] = c\left[\dfrac{2}{3}\left(\dfrac{1}{\theta}\right) + \dfrac{4}{3}\left(\dfrac{1}{\theta}\right)\right] = 2c\left(\dfrac{1}{\theta}\right).$ For the estimator to be

unbiased, $2c = 1$ or $c = 1/2$.

5.4.11 $E(W^2) = \text{Var}(W) + E(W)^2 = \text{Var}(W) + \theta^2$. Thus, W^2 is unbiased only if $\text{Var}(W) = 0$, which in essence means that $W = 0$.

5.4.13 $f_{\frac{n+1}{n}Y_{max}}(y) = \dfrac{n}{n+1} f_{Y_{max}}\left(\dfrac{n}{n+1}y\right) = \dfrac{n}{n+1}\dfrac{n}{\theta}\dfrac{n^{n-1}}{(n+1)^{n-1}}\dfrac{y^{n-1}}{\theta^{n-1}} = \dfrac{n^{n+1}}{(n+1)^n}\dfrac{y^{n-1}}{\theta^n}$

The median of this distribution is the number m such that

$$1/2 = \int_{0}^{m} \dfrac{n^{n+1}}{(n+1)^n}\dfrac{y^{n-1}}{\theta^n} dy = \dfrac{n^n}{(n+1)^n}\dfrac{y^n}{\theta^n}\Big|_0^m = \dfrac{n^n}{(n+1)^n}\dfrac{m^n}{\theta^n}$$

Solving for m gives $m = \dfrac{1}{\sqrt[n]{2}}\dfrac{(n+1)}{n}\theta$. The estimator is unbiased only when $n = 1$.

5.4.15 $E(\overline{W}^2) = \text{Var}(\overline{W}) + E(\overline{W})^2 = \dfrac{\sigma^2}{n} + \mu^2$, so $\lim_{n\to\infty} E(\overline{W}^2) = \lim_{n\to\infty}\left(\dfrac{\sigma^2}{n} + \mu^2\right) = \mu^2$

5.4.17 (a) $E(\hat{p}_i) = E(X_1) = p$, since X_1 is binomial with $n = 1$. $E(\hat{p}_2) = E\left(\dfrac{X}{n}\right) = \dfrac{1}{n}np = p$, since X is binomial.

(b) $\mathrm{Var}(\hat{p}_1) = p(1-p);\ \mathrm{Var}(X) = np(1-p)$. Then $\mathrm{Var}(\hat{p}_2) = \dfrac{np(1-p)}{n^2} = \dfrac{p(1-p)}{n}$, which is smaller than $\mathrm{Var}(\hat{p}_1)$ by a factor of n.

5.4.19 (a) See Question 5.4.14.

(b) $\mathrm{Var}(Y_1) = \theta^2$, since Y_1 is a gamma variable with parameters 1 and $1/\theta$.
$\mathrm{Var}(\bar{Y}) = \mathrm{Var}(Y_1)/n = \theta^2/n$
From Question 5.4.12, $\mathrm{Var}(nY_{\min}) = \mathrm{Var}(Y) = \theta^2$

(c) $\mathrm{Var}(\hat{\theta}_3)/\mathrm{Var}(\hat{\theta}_1) = \theta^2/\theta^2 = 1$
$\mathrm{Var}(\hat{\theta}_3)/\mathrm{Var}(\hat{\theta}_2) = \theta^2/(\theta^2/n) = n$

5.4.21 From Example 5.4.7, $\mathrm{Var}(\hat{\theta}_1) = \mathrm{Var}\left(\left(\dfrac{n+1}{n}\right)Y_{\max}\right) = \dfrac{\theta^2}{n(n+2)}$

Generalizing Question 5.4.18, we obtain

$$\mathrm{Var}(\hat{\theta}_2) = \mathrm{Var}\big((n+1)Y_{\min}\big) = \dfrac{n\theta^2}{(n+2)}$$

$$\mathrm{Var}(\hat{\theta}_2)/\mathrm{Var}(\hat{\theta}_1) = \dfrac{n\theta^2}{(n+2)}\Big/\dfrac{\theta^2}{n(n+2)} = n^2$$

Section 5.5

5.5.1 $E(\hat{\theta}) = \dfrac{3}{2}E(\bar{Y}) = \dfrac{3}{2}E(Y) = \dfrac{3}{2}\left(\dfrac{2}{3}\theta\right) = \theta$

5.5.3 $\ln f_X(X;\lambda) = -\lambda + X\ln \lambda - \ln X!$
$\dfrac{\partial \ln f_X(X;\lambda)}{\partial \lambda} = -1 + X/\lambda$
$\dfrac{\partial^2 \ln f_X(X;\lambda)}{\partial \lambda^2} = -X/\lambda^2$
$E\left[\dfrac{\partial^2 \ln f_X(X;\lambda)}{\partial \lambda^2}\right] = -\lambda/\lambda^2 = -1/\lambda$, so the Cramer-Rao bound is λ/n. Also, $\mathrm{Var}(\hat{\lambda}) = \mathrm{Var}(\bar{X})$
$= \mathrm{Var}(X)/n = \lambda/n$, so $\hat{\theta}$ is an efficient estimator.

5.5.5 $\ln f_Y(Y;\theta) = -\ln \theta$

$$\frac{\partial \ln f_Y(Y;\theta)}{\partial \theta} = \frac{-1}{\theta}$$

$$E\left[\left(\frac{\partial \ln f_Y(Y;\theta)}{\partial \theta}\right)^2\right] = \frac{1}{\theta^2}, \text{ so the Cramer-Rao bound is } \frac{\theta^2}{n}. \text{ From Question 5.4.21,}$$

$$\text{Var}(\hat{\theta}) = \frac{\theta^2}{n(n+2)}, \text{ which is smaller than the Cramer-Rao bound. This occurs because}$$

Theorem 5.5.1 is not necessarily valid if the range of the pdf depends on the parameter.

5.5.7 $$E\left(\frac{\partial^2 \ln f_W(W;\theta)}{\partial \theta^2}\right) = \int_{-\infty}^{\infty} \frac{\partial}{\partial \theta}\left(\frac{\partial \ln f_W(w;\theta)}{\partial \theta}\right) f_W(w;\theta)\, dw$$

$$= \int_{-\infty}^{\infty} \frac{\partial}{\partial \theta}\left(\frac{1}{f_W(w;\theta)} \frac{\partial f_W(w;\theta)}{\partial \theta}\right) f_W(w;\theta)\, dw$$

$$= \int_{-\infty}^{\infty} \left[\frac{1}{f_W(w;\theta)}\frac{\partial^2 f_W(w;\theta)}{\partial \theta^2} - \frac{1}{(f_W(w;\theta))^2}\left(\frac{\partial f_W(w;\theta)}{\partial \theta}\right)^2\right] f_W(w;\theta)\, dw$$

$$= \int_{-\infty}^{\infty} \frac{\partial^2 f_W(w;\theta)}{\partial \theta^2}\, dw - \int_{-\infty}^{\infty} \frac{1}{(f_W(w;\theta))^2}\left(\frac{\partial f_W(w;\theta)}{\partial \theta}\right)^2 f_W(w;\theta)\, dw$$

$$= 0 - \int_{-\infty}^{\infty}\left(\frac{\partial \ln f_W(w;\theta)}{\partial \theta}\right)^2 f_W(w;\theta)\, dw$$

The 0 occurs because $1 = \int_{-\infty}^{\infty} f_W(w;\theta)\, dw$, so

$$0 = \frac{\partial^2 \int_{-\infty}^{\infty} f_W(w;\theta)\, dw}{\partial \theta^2} = \int_{-\infty}^{\infty} \frac{\partial^2 f_W(w;\theta)}{\partial \theta^2}\, dw$$

The above argument shows that

$$E\left(\frac{\partial^2 \ln f_W(W;\theta)}{\partial \theta^2}\right) = -E\left(\frac{\partial \ln f_W(W;\theta)}{\partial \theta}\right)^2$$

Multiplying both sides of the equality by n and inverting gives the desired equality.

Section 5.6

5.6.1 $$\prod_{i=1}^{n} p_X(x_i;p) = \prod_{i=1}^{n}(1-p)^{x_i-1} p = (1-p)^{\left(\sum_{i=1}^{n} x_i\right)-n} p^n$$

Let $g\left(\sum_{i=1}^{n} x_i; p\right) = (1-p)^{\left(\sum_{i=1}^{n} x_i\right)-n} p^n$ and $u(x_1, \ldots x_n) = 1$.

By Theorem 5.6.1, the statistic $\sum_{i=1}^{n} X_i$ is sufficient

Chapter 5

61

5.6.3 $P((1, 1, 0) \mid X_1 + 2X_2 + 3X_3 = 3)$

$$= \frac{P((1, 1, 0) \text{ and } X_1 + 2X_2 + 3X_3 = 3)}{P(X_1 + 2X_2 + 3X_3 = 3)}$$

$$= \frac{P((1, 1, 0))}{P((1, 1, 0), (0, 0, 1))} = \frac{p^2(1-p)}{p^2(1-p) + p(1-p)^2} = p \, .$$

Since the conditional probability does depend on the parameter p, the statistic cannot be sufficient, by Definition 5.6.1.

5.6.5 $\displaystyle\prod_{i=1}^{n} \frac{1}{\sqrt{2\pi}\sigma} e^{-\frac{1}{2}\sum_{i=1}^{n}\frac{y_i^2}{\sigma^2}} = \left[(\sigma^2)^{-n/2} e^{-\frac{1}{2}\frac{1}{\sigma^2}\left(\sum_{i=1}^{n}y_i^2\right)} \right] [2\pi^{-n/2}],$ so $\displaystyle\sum_{i=1}^{n} Y_i^2$ is

sufficient by Theorem 5.6.1.

5.6.7 $L = \displaystyle\prod_{i=1}^{n} f_Y(y_i; \theta) = \prod_{i=1}^{n} \theta y_i^{\theta-1} = \theta^n \left(\prod_{i=1}^{n} y_i \right)^{\theta-1}$, and $\ln L = n \cdot \ln \theta + (\theta - 1) \displaystyle\sum_{i=1}^{n} \ln y_i$

$$\frac{d \ln L}{d\theta} = \frac{n}{\theta} + \sum_{i=1}^{n} \ln y_i$$

Setting $\dfrac{d \ln L}{d\theta} = 0$ gives $\theta_e = \dfrac{-n}{\displaystyle\sum_{i=1}^{n} \ln y_i} = \dfrac{-n}{\ln\left(\displaystyle\prod_{i=1}^{n} y_i \right)}$, which is a function of $\displaystyle\prod_{i=1}^{n} y_i$.

5.6.9 $\lambda e^{-\lambda y} = e^{\ln \lambda - \lambda y} = e^{y(-\lambda) + \ln \lambda}$. Take $K(y) = y$, $p(\lambda) = -\lambda$, $S(y) = 0$, and $q(\lambda) = \ln \lambda$.

Then $\displaystyle\sum_{i=1}^{n} Y_i$ is sufficient.

Section 5.7

5.7.1 $P(16 < \bar{Y} < 20) = 0.90$ is equivalent to $P\left(\dfrac{16-18}{5.0/\sqrt{n}} < Z < \dfrac{20-18}{5.0/\sqrt{n}} \right) = 0.90$ or

$P(-0.40\sqrt{n} < Z < 0.40\sqrt{n}) = 0.90.$ Then $0.40\sqrt{n} = 1.64$ or $n = \left(\dfrac{1.64}{0.40} \right)^2 = 16.81,$

so take $n = 17$.

5.7.3 (a) $P(Y_1 > 2\lambda) = \displaystyle\int_{2\lambda}^{\infty} \lambda e^{-\lambda y} dy = e^{-2\lambda^2}$. Then $P(\mid Y_1 - \lambda \mid < \lambda/2) < 1 - e^{-2\lambda^2} < 1.$

Thus, $\displaystyle\lim_{n \to \infty} P(\mid Y_1 - \lambda \mid) < \lambda/2) < 1.$

(b) $P\left(\displaystyle\sum_{i=1}^{n} Y_i > 2\lambda \right) \geq P(Y_1 > 2\lambda) = e^{-2\lambda^2}$. The proof now proceeds along the lines of Part

(a).

5.7.5 $E\left[(Y_{\max} - \theta)^2\right] = \int_0^\theta (y-\theta)^2 \frac{n}{\theta}\left(\frac{y}{\theta}\right)^{n-1} dy$

$$= \frac{n}{\theta^n}\int_0^\theta (y^{n+1} - 2\theta y^n + \theta^2 y^{n-1})dy = \frac{n}{\theta^n}\left(\frac{\theta^{n+2}}{n+2} - \frac{2\theta^{n+2}}{n+1} + \frac{\theta^{n+2}}{n}\right)$$

$$= \left(\frac{n}{n+2} - \frac{2n}{n+1} + 1\right)\theta^2$$

Then $\displaystyle\lim_{n\to\infty} E\left[(Y_{\max} - \theta)^2\right] = \lim_{n\to\infty}\left(\frac{n}{n+2} - \frac{2n}{n+1} + 1\right)\theta^2 = 0$ and the estimator is squared error consistent.

Section 5.8

5.8.1   The numerator of $g_\Theta(\theta \mid X = k)$ is

$$p_X(k \mid \theta)f_\Theta(\theta) = [(1-\theta)^k\theta]\frac{\Gamma(r+s)}{\Gamma(r)\Gamma(s)}\theta^{r-1}(1-\theta)^{s-1} = \frac{\Gamma(r+s)}{\Gamma(r)\Gamma(s)}\theta^r(1-\theta)^{s+k-1}$$

We recognize the part involving θ as the variable part of the beta distribution with parameters $r + 1$ and $s + k$, so that is $g_\Theta(\theta \mid X = k)$.

5.8.3   (a)   Following the pattern of Example 5.8.2, we can see that the posterior distribution is a beta pdf with parameters $k + 135$ and $n - k + 135$.

(b)   The mean of the Bayes pdf given in part (a) is $\dfrac{k+135}{k+135+n-k+135} = \dfrac{k+135}{n+270}$,

Note we can write

$$\frac{k+135}{n+270} = \frac{n}{n+270}\left(\frac{k}{n}\right) + \frac{270}{n+270}\left(\frac{135}{270}\right) = \frac{n}{n+270}\left(\frac{k}{n}\right) + \frac{270}{n+270}\left(\frac{1}{2}\right)$$

5.8.5   In each case the estimator is biased, since the mean of the estimator is a weighted average of the unbiased maximum likelihood estimator and a non-zero constant. However, in each case, the weighting on the maximum likelihood estimator tends to 1 as n tends to ∞, so these estimators are asymptotically unbiased.

5.8.7   Since the sum of gamma random variables is gamma, then W is gamma with parameters nr and λ. Then $g_\Theta(\theta \mid X = k)$ is a gamma pdf with parameters $nr + s$ and $\displaystyle\sum_{i=1}^n y_i + \mu$.

5.8.9 $p_X(k \mid \theta)f_\Theta(\theta) = \binom{n}{k}\dfrac{\Gamma(r+s)}{\Gamma(r)\Gamma(s)}\theta^{k+r-1}(1-\theta)^{n-k+s-1}$, so

$$p_X(k \mid \theta) = \binom{n}{k}\frac{\Gamma(r+s)}{\Gamma(r)\Gamma(s)}\int_0^1 \theta^{k+r-1}(1-\theta)^{n-k+s-1}d\theta,$$

$$= \binom{n}{k}\frac{\Gamma(r+s)}{\Gamma(r)\Gamma(s)}\frac{\Gamma(k+r)\Gamma(n-k+s)}{\Gamma(n+r+s)} = \frac{n!}{k!(n-k)!}\frac{(r+s-1)!}{(r-1)!(s-1)!}\frac{(k+r-1)!(n-k+s-1)!}{(n+r+s-1)!}$$

$$= \frac{(k+r-1)!}{k!(r-1)!}\frac{(n-k+s-1)!}{(n-k)!(s-1)!}\frac{n!(r+s-1)!}{(n+r+s-1)!}$$

$$= \binom{k+r-1}{k}\binom{n-k+s-1}{n-k} \Big/ \binom{n+r+s-1}{n}$$

Chapter 6

Section 6.2

6.2.1 (a) Reject H_0 if $\dfrac{\bar{y}-120}{18/\sqrt{25}} \le -1.41$; $z = -1.61$; reject H_0.

(b) Reject H_0 if $\dfrac{\bar{y}-42.9}{3.2/\sqrt{16}}$ is either 1) ≤ -2.58 or 2) ≥ 2.58; $z = 2.75$; reject H_0.

(c) Reject H_0 if $\dfrac{\bar{y}-14.2}{4.1/\sqrt{9}} \ge 1.13$; $z = 1.17$; reject H_0.

6.2.3 (a) No, because the observed z could fall <u>between</u> the 0.05 and 0.01 cutoffs.

(b) Yes. If the observed z exceeded the 0.01 cutoff, it would necessarily exceed the 0.05 cutoff.

6.2.5 No, because two-sided cutoffs (for a given α) are further away from 0 than one-sided cutoffs.

6.2.7 (a) H_0 should be rejected if $\dfrac{\bar{y}-12.6}{0.4/\sqrt{30}}$ is either 1) ≤ -1.96 or 2) ≥ 1.96. But $\bar{y} = 12.76$ and $z = 2.19$, suggesting that the machine should be readjusted.

(b) The test assumes that the y_i's constitute a random sample from a normal distribution. Graphed, a histogram of the 30 y_i's shows a mostly bell-shaped pattern. There is no reason to suspect that the normality assumption is not being met.

6.2.9 P-value $= P(Z \le -0.92) + P(Z \ge 0.92) = 0.3576$; H_0 would be rejected if α had been set at any value greater than or equal to 0.3576.

6.2.11 H_0 should be rejected if $\dfrac{\bar{y}-145.75}{9.50/\sqrt{25}}$ is either 1) ≤ -1.96 or 2) ≥ 1.96. Here, $\bar{y} = 149.75$ and $z = 2.10$, so the difference between \$145.75 and \$149.75 <u>is</u> statistically significant.

Section 6.3

6.3.1 (a) Given that the technique worked $k = 24$ times during the $n = 52$ occasions it was tried, $z = \dfrac{24 - 52(0.40)}{\sqrt{52(0.40)(0.60)}} = 0.91$. The latter is not larger than $z_{.05} = 1.64$, so $H_0\colon p = 0.40$ would not be rejected at the $\alpha = 0.05$ level. These data do not provide convincing evidence that transmitting predator sounds helps to reduce the number of whales in fishing waters.

(b) P-value $= P(Z \ge 0.91) = 0.1814$; H_0 would be rejected for any $\alpha \ge 0.1814$.

6.3.3   Let $p = P$(current supporter is male). Test H_0: $p = 0.65$ versus H_1: $p < 0.65$. Since $n = 120$ and $k =$ number of male supporters $= 72$, $z = \dfrac{72 - 120(0.65)}{\sqrt{120(0.65)(0.35)}} = -1.15$, which is not less than or equal to $-z_{.05}$ $(= -1.64)$, so H_0: $p = 0.65$ would not be rejected.

6.3.5   Let $p = P(Y_i \le 0.69315)$. Test H_0: $p = \dfrac{1}{2}$ versus H_1: $p \ne \dfrac{1}{2}$. Given that $k = 26$ and $n = 60$, P-value $= P(X \le 26) + P(X \ge 34) = 0.3030$.

6.3.7   Reject H_0 if $k \ge 4$ gives $\alpha = 0.50$; reject H_0 if $k \ge 5$ gives $\alpha = 0.23$; reject H_0 if $k \ge 6$ gives $\alpha = 0.06$; reject H_0 if $k \ge 7$ gives $\alpha = 0.01$.

6.3.9   (a) $\alpha = P(\text{reject } H_0 \mid H_0 \text{ is true}) = P(X \le 3 \mid p = 0.75) =$

$$\sum_{k=0}^{3} \binom{7}{k}(0.75)^k (0.25)^{7-k} = 0.07$$

(b)

p	$P(X \le 3 \mid p)$
0.75	0.07
0.65	0.20
0.55	0.39
0.45	0.61
0.35	0.80
0.25	0.93
0.15	0.99

Section 6.4

6.4.1   (a)   As described in Example 6.2.1, H_0: $\mu = 494$ is to be tested against H_1: $\mu \ne 494$ using ± 1.96 as the $\alpha = 0.05$ cutoffs. That is, H_0 is rejected if $\dfrac{\bar{y} - 494}{124/\sqrt{86}} \le -1.96$ or if $\dfrac{\bar{y} - 494}{124/\sqrt{86}} \ge 1.96$. Equivalently, the null hypothesis is rejected if $\bar{y} \le 467.8$ or if $\bar{y} \ge 520.2$. Therefore,

$1 - \beta = P(\text{reject } H_0 \mid \mu = 500) = P(\bar{Y} \le 467.8 \mid \mu = 500) + P(\bar{Y} \ge 520.2 \mid \mu = 500) =$

$P\left(Z \le \dfrac{467.8 - 500}{124/\sqrt{86}}\right) + P\left(Z \ge \dfrac{520.2 - 500}{124/\sqrt{86}}\right) = P(Z \le -2.41) + P(Z \ge 1.51) =$

$0.0080 + 0.0655 = 0.0735$.

6.4.3   The null hypothesis in Question 6.2.2 is rejected if $\bar{y}$ is either **1)** ≤ 89.0 or **2)** ≥ 101.0. Suppose $\mu = 90$. Since $\sigma = 15$ and $n = 22$, $1 - \beta = P(\bar{Y} \le 89.0) + P(\bar{Y} \ge 101.0) =$

$P\left(Z \le \dfrac{89.0 - 90}{15/\sqrt{22}}\right) + P\left(Z \ge \dfrac{101.0 - 90}{15/\sqrt{22}}\right) = P(Z \le -0.31) + P(Z \ge 3.44) = 0.3783 + 0.0003 =$

0.3786.

6.4.5 H_0 should be rejected if $z = \dfrac{\bar{y} - 240}{50/\sqrt{25}} \leq -2.33$ or , equivalently, if $\bar{y} \leq 240 - 2.33 \cdot \dfrac{50}{\sqrt{25}} = $
216.7. Suppose $\mu = 220$. Then $\beta = P(\text{accept } H_0 \mid H_1 \text{ is true}) = P(\bar{Y} > 216.7 \mid \mu = 220) = $
$P\left(Z > \dfrac{216.7 - 220}{50/\sqrt{25}} \right) = P(Z > -0.33) = 0.6293.$

6.4.7 For $\alpha = 0.10$, H_0: $\mu = 200$ should be rejected if $\bar{y} \leq 200 - 1.28 \cdot \dfrac{15.0}{\sqrt{n}}$. Also, $1 - \beta = $
$P\left(\bar{Y} \leq 200 - 1.28 \cdot \dfrac{15.0}{\sqrt{n}} \middle| \mu = 197 \right) = 0.75$, so $P\left(\dfrac{200 - 1.28 \cdot 15.0/\sqrt{n} - 197}{15.0/\sqrt{n}} \right) = 0.75$. But
$P(Z \leq 0.67) = 0.75$, implying that $\dfrac{200 - 1.28 \cdot 15.0/\sqrt{n} - 197}{15.0/\sqrt{n}} = 0.67$. It follows that the
smallest n satisfying the conditions placed on α and $1 - \beta$ is 95.

6.4.9 Since H_1 is one-sided, H_0 is rejected when $\bar{y} \geq 30 + z_\alpha \cdot \dfrac{9}{\sqrt{16}}$. Also, $1 - \beta = $ power $=$
$P\left(\bar{Y} \geq 30 + z_\alpha \cdot \dfrac{9}{\sqrt{16}} \middle| \mu = 34 \right) = 0.85$. Therefore, $1 - \beta = P\left(Z \geq \dfrac{30 + z_\alpha \cdot 9/\sqrt{16} - 34}{9/\sqrt{16}} \right) = $
0.85. But $P(Z \geq -1.04) = 0.85$, so $\dfrac{30 + z_\alpha \cdot 9/\sqrt{16} - 34}{9/\sqrt{16}} = -1.04$, implying that $z_\alpha = 0.74$.
Therefore, $\alpha = 0.23$.

6.4.11 In this context, α is the proportion of incorrect decisions made on innocent suspects—that is,
$\dfrac{9}{140}$, or 0.064. Similarly, β is the proportion of incorrect decisions made on guilty
suspects— here, $\dfrac{15}{140}$, or 0.107. A Type I error (convicting an innocent defendant) would
be considered more serious than a Type II error (acquitting a guilty defendant).

6.4.13 For a uniform pdf, $f_{Y_{max}}(y) = \dfrac{5}{\theta}\left(\dfrac{y}{\theta} \right)^4$, $0 \leq y \leq \theta$ when $n = 5$. Therefore,
$\alpha = P(\text{reject } H_0 \mid H_0 \text{ is true}) = P(Y_{max} \geq k \mid \theta = 2) = \int_k^2 \dfrac{5}{2}\left(\dfrac{y}{2} \right)^4 dy = 1 - \dfrac{k^5}{32}$. For α to be 0.05,
$k = 1.98$.

6.4.15 $\beta = P(\text{accept } H_0 \mid H_1 \text{ is true}) = P(X \leq n - 1 \mid p) = 1 - P(X = n \mid p) = 1 - \binom{n}{n} p^n (1-p)^0 = $
$1 - p^n$. When $\beta = 0.05$, $p = \sqrt[n]{0.95}$.

6.4.17 $1 - \beta = P(\text{reject } H_0 \mid H_1 \text{ is true}) = P\left(Y \leq \frac{1}{2}\Big| \theta\right) = \int_0^{1/2} (1+\theta)y^\theta dy = y^{\theta+1}\Big|_0^{1/2} = \left(\frac{1}{2}\right)^{\theta+1}.$

6.4.19 $P(\text{Type II error}) = \beta = P(\text{accept } H_0 \mid H_1 \text{ is true}) = P\left(X \leq 3 \Big| p = \frac{1}{2}\right) = \sum_{k=1}^{3}\left(1 - \frac{1}{2}\right)^{k-1} \cdot \frac{1}{2} = \frac{7}{8}.$

6.4.21 $\alpha = P(\text{reject } H_0 \mid H_0 \text{ is true}) = P(Y_1 + Y_2 \leq k \mid \theta = 2).$ When H_0 is true, Y_1 and Y_2 are uniformly distributed over the square defined by $0 \leq Y_1 \leq 2$ and $0 \leq Y_2 \leq 2$, so the joint pdf of Y_1 and Y_2 is a plane parallel to the Y_1Y_2-axis at height $\frac{1}{4}\left(= f_{Y_1}(y_1) \cdot f_{Y_2}(y_2) = \frac{1}{2} \cdot \frac{1}{2}\right)$. By geometry, α is the volume of the triangular wedge in the lower left-hand corner of the square over which Y_1 and Y_2 are defined. The hypotenuse of the triangle in the Y_1Y_2-plane has the equation $y_1 + y_2 = k$. Therefore, $\alpha = \text{area of triangle} \times \text{height of wedge} = \frac{1}{2} \cdot k \cdot k \cdot \frac{1}{4} = k^2/8.$ For α to be 0.05, $k = \sqrt{0.04} = 0.63.$

Section 6.5

6.5.1 $L(\hat{\omega}) = \prod_{i=1}^{n} (1-p_0)^{k_i-1} p_0 = p_0^n (1-p_0)^{\sum_{i=1}^{n} k_i - n} = p_0^n (1-p_0)^{k-n}$, where $k = \sum_{i=1}^{n} k_i$. From the comment following Example 5.2.1, the maximum likelihood estimate for p is $p_e = \frac{n}{k}$.

Therefore, $L(\hat{\Omega}) = \left(\frac{n}{k}\right)^n \left(1 - \frac{n}{k}\right)^{k-n}$, and the generalized likelihood ratio for testing H_0: $p = p_0$ versus H_1: $p \neq p_0$ is the quotient $L(\hat{\omega})/L(\hat{\Omega})$.

6.5.3 $L(\hat{\omega}) = \prod_{i=1}^{n} (1/\sqrt{2\pi})e^{-\frac{1}{2}(y_i - \mu_0)^2} = (2\pi)^{-n/2} e^{-\frac{1}{2}\sum_{i=1}^{n}(y_i - \mu_0)^2}$. Since $\bar{y}$ is the maximum likelihood estimate for μ (recall the first derivative taken in Example 5.2.4),

$L(\hat{\Omega}) = (2\pi)^{-n/2} e^{-\frac{1}{2}\sum_{i=1}^{n}(y_i - \bar{y})^2}$. Here the generalized likelihood ratio reduces to

$\lambda = L(\hat{\omega})/L(\hat{\Omega}) = e^{-\frac{1}{2}((\bar{y}-\mu_0)/(1/\sqrt{n}))^2}$. The null hypothesis should be rejected if

$e^{-\frac{1}{2}((\bar{y}-\mu_0)/(1/\sqrt{n}))^2} \leq \lambda^*$ or, equivalently, if $\left|(\bar{y} - \mu_0)\right|/(1/\sqrt{n}) > \lambda^{**}$, where values for λ^{**} come from the standard normal pdf, $f_Z(z)$.

6.5.5   (a) $\lambda = \left(\dfrac{1}{2}\right)^{n} / [(x/n)^{x}(1 - x/n)^{n-x}] = 2^{-n}x^{-x}(n-x)^{x-n}n^{n}$. Rejecting H_0 when $0 < \lambda \le \lambda^*$

is equivalent to rejecting H_0 when $x\ln x + (n-x)\ln(n-x) \ge \lambda^{**}$.

(b)   By inspection, $x\ln x + (n-x)\ln(n-x)$ is symmetric in x. Therefore, the left-tail and

right-tail critical regions will be equidistant from $p = \dfrac{1}{2}$, which implies that H_0 should

be rejected if $\left| x - \dfrac{1}{2} \right| \ge k$, where k is a function of α.

Chapter 7

Section 7.3

7.3.1 Clearly, $f_Y(y) > 0$ for all $y > 0$. To verify that $f_Y(y)$ is a pdf requires proving that $\int_0^\infty f_Y(y)dy = 1$.

But $\int_0^\infty f_Y(y)dy = \dfrac{1}{\Gamma(n/2)}\int_0^\infty \dfrac{1}{2^{n/2}}y^{n/2\,-\,1}e^{-y/2}dy$. By definition, $\Gamma\left(\dfrac{n}{2}\right) = \int_0^\infty u^{n/2\,-\,1}e^{-u}du$.

Let $u = \dfrac{y}{2}$, so $du = \dfrac{dy}{2}$. Then $\Gamma\left(\dfrac{n}{2}\right) = \int_0^\infty \left(\dfrac{y}{2}\right)^{n/2\,-\,1}e^{-y/2}dy/2 = \int_0^\infty \dfrac{1}{2^{n/2}}y^{n/2\,-\,1}e^{-y/2}dy$.

Therefore, $\int_0^\infty f_Y(y)dy = \dfrac{1}{\Gamma(n/2)}\cdot\Gamma\left(\dfrac{n}{2}\right) = 1$.

7.3.3 If $\mu = 50$ and $\sigma = 10$, $\sum_{i=1}^{3}\left(\dfrac{Y_i - 50}{10}\right)^2$ should have a χ_3^2 distribution, implying that the numerical value of the sum is likely to be between, say, $\chi_{.025,3}^2$ ($=0.216$) and $\chi_{.975,3}^2$ ($=9.348$). Here, $\sum_{i=1}^{3}\left(\dfrac{Y_i - 50}{10}\right)^2 = \left(\dfrac{65-50}{10}\right)^2 + \left(\dfrac{30-50}{10}\right)^2 + \left(\dfrac{55-50}{10}\right)^2 = 6.50$, so the data are not inconsistent with the hypothesis that the Y_i's are normally distributed with $\mu = 50$ and $\sigma = 10$.

7.3.5 Since $E(S^2) = \sigma^2$, it follows from Chebyshev's inequality that $P(|S^2 - \sigma^2| < \varepsilon) > 1 - \dfrac{\text{Var}(S^2)}{\varepsilon^2}$. But $\text{Var}(S^2) = \dfrac{2\sigma^4}{n-1} \to 0$ as $n \to \infty$. Therefore, S^2 is consistent for σ^2.

7.3.7 (a) 0.983   (b) 0.132   (c) 9.00

7.3.9 (a) 6.23   (b) 0.65   (c) 9   (d) 15
      (e) 2.28

7.3.11 $F = \dfrac{V/m}{U/n}$, where U and V are independent χ^2 random variables with m and n degrees of freedom, respectively. Then $\dfrac{1}{F} = \dfrac{U/n}{V/m}$, which implies that $\dfrac{1}{F}$ has an F distribution with n and m degrees of freedom.

7.4.13 To show that $f_{T_n}(t)$ converges to $f_Z(t)$ requires proving that **1)** $\left(1 + \dfrac{t^2}{n}\right)^{-(n+1)/2}$ converges to

$e^{-t^2/2}$ and **2)** $\dfrac{\Gamma\left(\dfrac{n+1}{2}\right)}{\sqrt{n\pi}\,\Gamma\left(\dfrac{n}{2}\right)}$ converges to $1/\sqrt{2\pi}$. To verify **1)**, write $\left(1 + \dfrac{t^2}{n}\right)^{-(n+1)/2} =$

$\left\{\left(1 + \dfrac{1}{n/t^2}\right)^{n/t^2}\right\}^{-t^2/2} \cdot \left(1 + \dfrac{t^2}{n}\right)^{-1/2}$. As n gets large, the last factor approaches 1 and the

product approaches $(e^1)^{-t^2/2} = e^{-t^2/2}$. Also, for large n, $n! \doteq \sqrt{2\pi n}\, n^n e^{-n}$. Equivalently,

$\Gamma(r) \doteq \sqrt{\dfrac{2\pi}{r}}\left(\dfrac{r}{e}\right)^r$ if r is large. An application of the latter shows that $\Gamma\left(\dfrac{n+1}{2}\right)\Big/\Gamma\left(\dfrac{n}{2}\right)$

converges to $\sqrt{\dfrac{n}{2}}$, which means that the constant in $f_{T_n}(t)$ converges to $1/\sqrt{2\pi}$, the constant

in the standard normal pdf.

Section 7.4

7.4.1 (a) 0.15   (b) 0.80   (c) 0.85
          (d) $0.99 - 0.15 = 0.84$

7.4.3 Both differences represent intervals associated with 5% of the area under $f_{T_n}(t)$. Because the pdf is closer to the horizontal axis the further t is away from 0, the difference $t_{.05,n} - t_{.10,n}$ is the larger of the two.

7.4.5 $P\left(\left|\dfrac{\overline{Y} - 15.0}{S/\sqrt{11}}\right| \geq k\right) = 0.05$ implies that $P\left(-k \leq \dfrac{\overline{Y} - 15.0}{S/\sqrt{11}} \leq k\right) = 0.95$. But $\dfrac{\overline{Y} - 15.0}{S/\sqrt{11}}$ is a

Student t random variable with 10 df. From Appendix Table A.2, $P(-2.2281 \leq T_{10} \leq 2.2281) = 0.95$, so $k = 2.2281$.

7.4.7 Since $n = 10$, $t_{\alpha/2, n-1} = t_{.005,9} = 3.2498$. For these data $\displaystyle\sum_{i=1}^{n} y_i = 4.84$ and $\displaystyle\sum_{i=1}^{n} y_i^2 = 2.86$.

Then $\overline{y} = 4.84/10 = 0.484$ and $s = \sqrt{\dfrac{10(2.86) - (4.84)^2}{10(9)}} = 0.240$.

The confidence interval is

$\left(\overline{y} - t_{\alpha/2, n-1}\dfrac{s}{\sqrt{n}}, \overline{y} + t_{\alpha/2, n-1}\dfrac{s}{\sqrt{n}}\right) = \left(0.484 - 3.2498\dfrac{0.240}{\sqrt{10}}, 0.484 + 3.2498\dfrac{0.240}{\sqrt{10}}\right)$

$= (0.237, 0.731)$.

7.4.9 a) Let μ = true average age at which scientists make their greatest discoveries. Since

$$\sum_{i=1}^{12} y_i = 425 \text{ and } \sum_{y=1}^{12} y_i^2 = 15{,}627,\ \bar{y} = \frac{1}{12}(425) = 35.4 \text{ and}$$

$$s = \sqrt{\frac{12(15{,}627) - (425)^2}{12(11)}} = 7.2.\ \text{Also,}\ t_{\alpha/2,n-1} = t_{.025,11} = 2.2010,\ \text{so the 95\% confidence}$$

interval for μ is the range $\left(35.4 - 2.2010 \cdot \dfrac{7.2}{\sqrt{12}}, 35.4 + 2.2010 \cdot \dfrac{7.2}{\sqrt{12}}\right)$, or (30.8 yrs, 40.0 yrs).

b) The graph of date versus age shows no obvious patterns or trends. The assumption that μ has remained constant over time is believable.

7.4.11 For $n = 24$, $t_{\alpha/2,n-1} = t_{.05,23} = 1.7139$. For these data $\bar{y} = 4645/24 = 193.54$ and

$$s = \sqrt{\frac{24(959{,}265) - (4645)^2}{24(23)}} = 51.19.$$

The confidence interval is $\left(193.54 - 1.7139\dfrac{51.19}{\sqrt{24}}, 193.54 + 1.7139\dfrac{51.19}{\sqrt{24}}\right) = (175.6, 211.4)$.

The medical and statistical definition of "normal" differ somewhat. There are people with medically norm platelet counts who appear in the population less than 10% of the time.

7.4.13 No, because the length of a confidence interval for μ is a function of s as well as the confidence coefficient. If the sample standard deviation for the second sample was sufficiently small (relative to the sample standard deviation for the first sample), the 95% confidence interval would be shorter than the 90% confidence interval.

7.4.15 a) 0.95   b) 0.80   c) 0.945   d) 0.95

7.4.17 Let μ = true average FEV_1/VC ratio for exposed workers. Since $\sum_{i=1}^{19} y_i = 14.56$ and

$$\sum_{i=1}^{19} y_i^2 = 11.2904,\ \bar{y} = \frac{14.56}{19} = 0.766 \text{ and } s = \sqrt{\frac{19(11.2904) - (14.56)^2}{19(18)}} = 0.0859.\ \text{To test}$$

H_0: $\mu = 0.80$ versus H_1: $\mu < 0.80$ at the $\alpha = 0.05$ level of significance, reject the null hypothesis if $t \leq -t_{.05,18} = -1.7341$. But $t = \dfrac{0.766 - 0.80}{0.0859/\sqrt{19}} = -1.71$, so we fail to reject H_0.

7.4.19 Let μ = true average GMAT increase earned by students taking the review course. The hypotheses to be tested are H_0: $\mu = 40$ versus H_1: $\mu < 40$. Here, $\sum_{i=1}^{15} y_i = 556$ and

$$\sum_{i=1}^{15} y_i^2 = 20{,}966,\ \text{so}\ \bar{y} = \frac{556}{15} = 37.1, s = \sqrt{\frac{15(20{,}966) - (556)^2}{15(14)}} = 5.0, \text{ and } t = \frac{37.1 - 40}{5.0/\sqrt{15}} =$$

-2.25. Since $-t_{.05,14} = -1.7613$, H_0 should be rejected at the $\alpha = 0.05$ level of significance, suggesting that the MBAs 'R Us advertisement may be fraudulent.

7.4.21 Let u = true average pit depth associated with plastic coating. To test H_0: $\mu = 0.0042$ versus H_1: $\mu < 0.0042$ at the $\alpha = 0.05$ level, we should reject the null hypothesis if $t \leq -t_{.05,9} = -1.8331$. For the 10 y_i's, $\bar{y} = \dfrac{0.0390}{10} = 0.0039$. Also, $s = 0.00383$, so $t = \dfrac{0.0039 - 0.0042}{0.00383/\sqrt{10}} = -2.48$. Since H_0 is rejected, these data support the claim that the plastic coating is an effective corrosion retardant.

7.4.23 Because of the skewed shape of $f_Y(y)$, and if the sample size was small, it would not be unusual for all the y_i's to lie close together near 0. When that happens, $\bar{y}$ will be less than μ, s will be considerably smaller than $E(S)$, and the t ratio will be further to the left of 0 than $f_{T_{n-1}}(t)$ would predict.

7.4.25 As n increases, Student t pdfs converge to the standard normal, $f_Z(z)$ (see Question 7.3.13).

Section 7.5

7.5.1 (a) 23.685   (b) 4.605   (c) 2.700

7.5.3 (a) 2.088   (b) 7.261   (c) 14.041   (d) 17.539

7.5.5 $\chi^2_{.95,200} \doteq 200\left(1 - \dfrac{2}{9(200)} + 1.64\sqrt{\dfrac{2}{9(200)}}\right)^3 = 233.9$

7.5.7 $P\left(\chi^2_{\alpha/2,n-1} \leq \dfrac{(n-1)S^2}{\sigma^2} \leq \chi^2_{1-\alpha/2,n-1}\right) = 1 - \alpha = P\left(\dfrac{(n-1)S^2}{\chi^2_{1-\alpha/2,n-1}} \leq \sigma^2 \leq \dfrac{(n-1)S^2}{\chi^2_{\alpha/2,n-1}}\right)$, so

$\left(\dfrac{(n-1)s^2}{\chi^2_{1-\alpha/2,n-1}}, \dfrac{(n-1)s^2}{\chi^2_{\alpha/2,n-1}}\right)$ is a 100(1 − α)% confidence interval for σ^2. Taking the square root of both sides gives a 100(1 − α)% confidence interval for σ.

7.5.9 (a) $\displaystyle\sum_{i=1}^{18} y_i = 1447$, so $\bar{y} = \dfrac{1447}{18} = 80.4$ and $s = \sqrt{\dfrac{1}{17}\sum_{i=1}^{17}(y_i - 80.4)^2} = 5.1$. Since $\chi^2_{.025,17} = 7.564$ and $\chi^2_{.975,17} = 30.191$, a 95% confidence interval for σ is $\left(\sqrt{\dfrac{17(5.1)^2}{30.191}}, \sqrt{\dfrac{17(5.1)^2}{7.564}}\right)$, or (3.8, 7.6).

(b) Given that $\chi^2_{.05,17} = 8.762$ and $\chi^2_{.95,17} = 27.587$, the two one-sided confidence intervals for σ are $\left(-\infty, \sqrt{\dfrac{17(5.1)^2}{8.672}}\right) = (-\infty, 7.1)$ and $\left(\sqrt{\dfrac{17(5.1)^2}{27.587}}, \infty\right) = (4.0, \infty)$.

7.5.11 (a) If $\dfrac{\chi^2_{n-1}-(n-1)}{\sqrt{2(n-1)}} \doteq Z$, then $P\left(-z_{\alpha/2} \le \dfrac{\chi^2_{n-1}-(n-1)}{\sqrt{2(n-1)}} \le z_{\alpha/2}\right) \doteq 1-\alpha =$

$$P\left(n-1-z_{\alpha/2}\sqrt{2(n-1)} \le \frac{(n-1)S^2}{\sigma^2} \le n-1+z_{\alpha/2}\sqrt{2(n-1)}\right) =$$

$$P\left(\frac{(n-1)S^2}{n-1+z_{\alpha/2}\sqrt{2(n-1)}} \le \sigma^2 \le \frac{(n-1)S^2}{n-1-z_{\alpha/2}\sqrt{2(n-1)}}\right),\ \text{so} \left(\frac{(n-1)s^2}{n-1+z_{\alpha/2}\sqrt{2(n-1)}},\right.$$

$$\left.\frac{(n-1)s^2}{n-1-z_{\alpha/2}\sqrt{2(n-1)}}\right) \text{ is an approximate } 100(1-\alpha)\% \text{ confidence interval for } \sigma^2.$$

Likewise, $\left(\dfrac{\sqrt{n-1}\,s}{\sqrt{n-1+z_{\alpha/2}\sqrt{2(n-1)}}}, \dfrac{\sqrt{n-1}\,s}{\sqrt{n-1-z_{\alpha/2}\sqrt{2(n-1)}}}\right)$ is an approximate

$100(1-\alpha)\%$ confidence interval for σ.

(b) For the data in Table 7.5.1, $n=19$ and $s=\sqrt{733.4}=27.08$, so the formula in Part a

gives $\left(\dfrac{\sqrt{18}(27.08)}{\sqrt{18+1.96\sqrt{36}}}, \dfrac{\sqrt{18}(27.08)}{\sqrt{18-1.96\sqrt{36}}}\right) = (21.1$ million years, 46.0 million years) as

the approximate 95% confidence interval for σ.

7.5.13 (a) $M_Y(t) = \dfrac{1}{1-\theta t}$. Let $X = \dfrac{2n\bar{Y}}{\theta} = \dfrac{2\sum_{i=1}^{n} Y_i}{\theta}$. Then $M_X(t) = \prod_{i=1}^{n} M_{Y_i}\left(\dfrac{2t}{\theta}\right) = \left(\dfrac{1}{1-2t}\right)^{2n/2}$,

implying that X is a χ^2_{2n} random variable.

(b) $P\left(\chi^2_{\alpha/2,2n} \le \dfrac{2n\bar{Y}}{\theta} \le \chi^2_{1-\alpha/2,2n}\right) = 1-\alpha$, so $\left(\dfrac{2n\bar{y}}{\chi^2_{1-\alpha/2,2n}}, \dfrac{2n\bar{y}}{\chi^2_{\alpha/2,2n}}\right)$ is a

$100(1-\alpha)\%$ confidence interval for θ.

7.5.15 Test $H_0: \sigma^2 = 1$ versus $H_1: \sigma^2 > 1$

The sample variance is $\dfrac{30(19{,}195.7938)-(758.62)^2}{30(29)} = 0.425$

The test statistic is $\chi^2 = \dfrac{29(0.425)}{1} = 12.325$. The critical value is $\chi^2_{.95,29} = 42.557$.

Since $12.325 < 42.557$, we accept the null hypothesis and assume the machine is working properly.

Chapter 8

Section 8.2

8.2.35 If there is a reason to believe the data varies over time, it would be regression data; otherwise, it would be one-sample data.

Chapter 9

Section 9.2

9.2.1 $t = \dfrac{\bar{x} - \bar{y}}{s_p\sqrt{1/n + 1/m}} = \dfrac{65.2 - 75.5}{13.9\sqrt{1/9 + 1/12}} = -1.68$

Since $-t_{.05,19} = -1.7291 < t = -1.68$, accept H_0.

9.2.3 $s_p = \sqrt{\dfrac{(n-2)s_X^2 + (m-1)s_Y^2}{n + m - 2}} = \sqrt{\dfrac{5(167.568 + 11(52.072)}{6 + 12 - 2}} = 9.39.$

$t = \dfrac{\bar{x} - \bar{y}}{s_p\sqrt{1/n + 1/m}} = \dfrac{28.6 - 12.758}{9.39\sqrt{1/6 + 1/12}} = 3.37$

Since $t = 3.37 > 2.9208 = t_{.005,16}$, reject H_0.

9.2.5 $s_p = \sqrt{\dfrac{7(7.169) + 7(10.304)}{8 + 8 - 2}} = 2.956$

$t = \dfrac{11.2 - 9.7875}{2.956\sqrt{1/8 + 1/8}} = 0.96$

Since $t = 0.96 < t_{.05,14} = 1.7613$, accept H_0.

9.2.7 Let $\alpha = 0.10$. $s_p = \sqrt{\dfrac{99(600^2) + 49(700^2)}{100 + 50 - 2}} = 634.9$

$t = \dfrac{2000 - 2500}{634.9\sqrt{1/100 + 1/50}} = -4.55$

Since $t = -4.55 < -t_{.05,148} = -z_{.05} = -1.64$, reject H_0

9.2.9 (a) Reject H_0 if $t > t_{.005,15} = 2.9467$, so we seek the smallest value of $|\bar{x} - \bar{y}|$ such that

$t = \dfrac{|\bar{x} - \bar{y}|}{s_p\sqrt{1/n + 1/m}} = \dfrac{|\bar{x} - \bar{y}|}{15.3\sqrt{1/6 + 1/11}} > 2.9467$, or $|\bar{x} - \bar{y}| > (15.3)(0.508)(2.9467)$

$= 22.90$

(b) Reject H_0 if $t > t_{.05,19} = 1.7291$, so we seek the smallest value of $\bar{x} - \bar{y}$ such that

$t = \dfrac{\bar{x} - \bar{y}}{s_p\sqrt{1/n + 1/m}} = \dfrac{\bar{x} - \bar{y}}{214.9\sqrt{1/13 + 1/8}} > 1.7291$, or $\bar{x} - \bar{y} > (214.9)(0.44936)(1.7291)$

$= 166.97$

9.2.11 (a) Let X be the interstate route; Y, the town route.
$P(X > Y) = P(X - Y > 0)$. $\mathrm{Var}(X - Y) = \mathrm{Var}(X) + \mathrm{Var}(Y) = 6^2 + 5^2 = 61$.

$P(X - Y > 0) = P\left(\dfrac{X - Y - (33 - 35)}{\sqrt{61}} > \dfrac{2}{\sqrt{61}}\right)$

$= P(Z \geq 0.26) = 1 - 0.6026 = 0.3974$

(b) $\text{Var}(\bar{X} - \bar{Y}) = \text{Var}(\bar{X}) + \text{Var}(\bar{Y}) = 6^2/10 + 5^2/10 = 61/10$

$$P(\bar{X} - \bar{Y}) > 0 = P\left(\frac{\bar{X} - \bar{Y} - (33 - 35)}{\sqrt{61/10}} > \frac{2}{\sqrt{61/10}}\right)$$
$$= P(Z > 0.81) = 1 - 0.7910 = 0.2090$$

9.2.13 $E(S_X^2) = E(S_Y^2) = \sigma^2$ by Example 5.4.4.

$$E(S_p^2) = \frac{(n-1)E(S_X^2) + (m-1)E(S_Y^2)}{n+m-2}$$
$$= \frac{(n-1)\sigma^2 + (m-1)\sigma^2}{n+m-2} = \sigma^2$$

9.2.15 For the data given, $\bar{x} = 545.45$, $s_X = 428$, and $\bar{y} = 241.82$, $s_Y = 183$. Then

$$t = \frac{\bar{x} - \bar{y}}{\sqrt{s_X^2/n + s_Y^2/m}} = \frac{545.45 - 241.82}{\sqrt{428^2/11 + 183^2/11}} = 2.16$$

The degrees of freedom associated with this statistic is the greatest integer in

$$\frac{(s_X^2/n + s_Y^2/m)^2}{(s_X^2/n)^2(n-1) + (s_Y^2/m)^2/(m-1)} = \frac{(428^2/11 + 183^2/11)^2}{(428^2/11)^2/10 + (183^2/11)^2/10} = 13.5.$$

The greatest integer is 13. Since $t = 2.16 > t_{.05,13} = 1.7709$, reject H_0.

9.2.17 (a) The sample standard deviation for the first data set is approximately 3.15; for the second, 3.29. These seem close enough to permit the use of Theorem 9.2.2.

(b) Intuitively, the states with the comprehensive law should have fewer deaths. However, the average for these data is 8.1, which is larger than the average of 7.0 for the states with a more limited law.

Section 9.3

9.3.1 (a) The critical values are $F_{.025,25,4}$ and $F_{.975,25,4}$. These values are not tabulated, but in this case, we can approximate them by $F_{.025,24,4} = 0.296$ and $F_{.975,24,4} = 8.51$. The observed $F = 86.9/73.6 = 1.181$. Since $0.296 < 1.181 < 8.51$, we can accept H_0 that the variances are equal.

(b) Yes, we can use Theorem 9.2.2, since we have no reason to doubt that the variances are equal.

9.3.3 (a) The critical values are $F_{.025,19,19}$ and $F_{.975,19,19}$. These values are not tabulated, but in this case, we can approximate them by $F_{.025,20,20} = 0.406$ and $F_{.975,20,20} = 2.46$. The observed $F = 2.41/3.52 = 0.685$. Since $0.406 < 0.685 < 2.46$, we can accept H_0 that the variances are equal.

(b) Since $t = 2.662 > t_{.025,38} = 2.0244$, reject H_0.

9.3.5 $F = 0.20^2/0.37^2 = 0.292$. Since $F_{.025,9,9} = 0.248 < 0.292 < 4.03 = F_{.975,9,9}$, accept H_0.

9.3.7 Let $\alpha = 0.05$. $F = 65.25/227.77 = 0.286$. Since $0.208 = F_{.025,8,5} < 0.286 < 6.76 = F_{.975,8,5}$, accept H_0. Thus, Theorem 9.2.2 is appropriate.

9.3.9 If $\sigma_X^2 = \sigma_Y^2 = \sigma^2$, the maximum likelihood estimator for σ^2 is

$$\hat{\sigma}^2 = \frac{1}{n+m}\left(\sum_{i=1}^n (x_i - \bar{x})^2 + \sum_{i=1}^m (y_i - \bar{y})^2\right). \text{ Then } L(\hat{\omega}) = \left(\frac{1}{2\pi\hat{\sigma}^2}\right)^{(n+m)/2} e^{-\frac{1}{2\hat{\sigma}^2}\left(\sum_{i=1}^n (x_i - \bar{x})^2 + \sum_{i=1}^m (y_i - \bar{y})^2\right)} =$$

$$\left(\frac{1}{2\pi\hat{\sigma}^2}\right)^{(n+m)/2} e^{-(n+m)/2}$$

If $\sigma_X^2 \neq \sigma_Y^2$ the maximum likelihood estimators for σ_X^2 and σ_Y^2 are

$$\hat{\sigma}_X^2 = \frac{1}{n}\sum_{i=1}^n (x_i - \bar{x})^2, \text{ and } \hat{\sigma}_Y^2 = \frac{1}{m}\sum_{i=1}^m (y_i - \bar{y})^2.$$

Then $L(\hat{\Omega}) = \left(\frac{1}{2\pi\hat{\sigma}_X^2}\right)^{n/2} e^{-\frac{1}{2\hat{\sigma}_X^2}\left(\sum_{i=1}^n (x_i - \bar{x})^2\right)} \left(\frac{1}{2\pi\hat{\sigma}_Y^2}\right)^{m/2} e^{-\frac{1}{2\hat{\sigma}_Y^2}\left(\sum_{i=1}^m (y_i - \bar{y})^2\right)}$

$$= \left(\frac{1}{2\pi\hat{\sigma}_X^2}\right)^{n/2} e^{-m/2} \left(\frac{1}{2\pi\hat{\sigma}_Y^2}\right)^{m/2} e^{-n/2}$$

The ratio $\lambda = \dfrac{L(\hat{\omega})}{L(\hat{\Omega})} = \dfrac{(\hat{\sigma}_X^2)^{n/2}(\hat{\sigma}_Y^2)^{m/2}}{(\hat{\sigma}^2)^{(n+m)/2}}$, which equates to the expression given in the statement

of the question.

Section 9.4

9.4.1 $\hat{p} = \dfrac{x+y}{n+m} = \dfrac{55+40}{200+200} = 0.2375$

$$z = \frac{\dfrac{x}{n} - \dfrac{y}{m}}{\sqrt{\dfrac{\hat{p}(1-\hat{p})}{n} + \dfrac{\hat{p}(1-\hat{p})}{m}}} = \frac{\dfrac{55}{200} - \dfrac{40}{200}}{\sqrt{\dfrac{0.2375(0.7625)}{200} + \dfrac{0.2375(0.7625)}{200}}} = 1.76$$

Since $-1.96 < z = 1.76 < 1.96 = z_{.025,}$ accept H_0.

9.4.3 Let $\alpha = 0.05$. $\hat{p} = \dfrac{24+27}{29+32} = 0.836$

$$z = \frac{\dfrac{24}{29} - \dfrac{27}{32}}{\sqrt{\dfrac{0.836(0.164)}{29} + \dfrac{0.836(0.164)}{32}}} = -0.17$$

For this experiment, H_0: $p_X = p_Y$ and H_1: $p_X \neq p_Y$. Since $-1.96 < z = -0.17 < 1.96 = z_{.025}$, accept H_0 at the 0.05 level of significance.

9.4.5 $\hat{p} = \dfrac{60+48}{100+100} = 0.54$

$$z = \frac{\dfrac{60}{100} - \dfrac{48}{100}}{\sqrt{\dfrac{0.54(0.46)}{100} + \dfrac{0.54(0.46)}{100}}} = 1.70$$

The P value is $P(Z \leq -1.70) + P(Z \geq 1.70) = 2(1 - 0.9554) = 0.0892$.

9.4.7 $\hat{p} = \dfrac{175+100}{609+160} = 0.358$

$$z = \frac{\dfrac{175}{609} - \dfrac{100}{160}}{\sqrt{\dfrac{0.358(0.642)}{609} + \dfrac{0.358(0.642)}{160}}} = -7.93. \text{ Since } z = -7.93 < -1.96 = -z_{.025}, \text{ reject } H_0.$$

9.4.9 From Equation 9.4.1,

$$\lambda = \frac{[(55+60)/(160+192)]^{(55+60)}[1-(55+60)/(160+192)]^{(160+192-55-60)}}{(55/160)^{55}[1-(55/160)]^{105}(60/192)^{60}[1-(60/192)]^{132}}$$

$$= \frac{115^{115}(237^{237})(160^{160})(192^{192})}{352^{352}(55^{55})(105^{105})(60^{60})(132^{132})}. \text{ We calculate } \ln \lambda, \text{ which is } -0.1935. \text{ Then } -2\ln \lambda =$$

$0.387.$ Since $-2\ln \lambda = 0.387 < 6.635 = \chi^2_{.99,1}$, accept H_0.

Section 9.5

9.5.1 The center of the confidence interval is $\bar{x} - \bar{y} = 1007.9 - 831.9 = 176.0$. The radius is

$$t_{\alpha/2,n+m-2}S_p\sqrt{\frac{1}{n} + \frac{1}{m}} = 2.0739(411)\sqrt{\frac{1}{9} + \frac{1}{15}} = 359.4. \text{ The confidence interval is}$$

$(176.0 - 359.4, 176.0 + 359.4) = (-183.4, 535.4)$

9.5.3 The center of the confidence interval is $\bar{x} - \bar{y} = 83.96 - 84.84 = -0.88$. The radius is

$$t_{\alpha/2,n+m-2}S_p\sqrt{\frac{1}{n} + \frac{1}{m}} = 2.2281(11.2)\sqrt{\frac{1}{5} + \frac{1}{7}} = 14.61. \text{ The confidence interval is } (-0.88 - $$

$14.61, -0.88 + 14.61) = (-15.49, 13.73)$. Since the confidence interval contains 0, the data do not suggest that the dome makes a difference.

9.5.5 Equation (9.5.1) is $P\left(-t_{\alpha/2,n+m-2} \le \dfrac{\bar{X} - \bar{Y} - (\mu_X - \mu_Y)}{S_p\sqrt{\dfrac{1}{n} + \dfrac{1}{m}}} \le t_{\alpha/2,n+m-2} \right) = 1 - \alpha$

which implies

$$P\left(-t_{\alpha/2,n+m-2}S_p\sqrt{\frac{1}{n} + \frac{1}{m}} \le \bar{X} - \bar{Y} - (\mu_X - \mu_Y) \le t_{\alpha/2,n+m-2}S_p\sqrt{\frac{1}{n} + \frac{1}{m}} \right) = 1 - \alpha, \text{ or}$$

$$P\left(-(\bar{X} - \bar{Y}) - t_{\alpha/2,n+m-2}S_p\sqrt{\frac{1}{n} + \frac{1}{m}} \le -(\mu_X - \mu_Y) \le -(\bar{X} - \bar{Y}) + t_{\alpha/2,n+m-2}S_p\sqrt{\frac{1}{n} + \frac{1}{m}} \right)$$

$= 1 - \alpha$

Multiplying the inequality above by -1 gives the inequality of the confidence interval of Theorem 9.5.1.

9.5.7 Approximate the needed $F_{.025,25,4}$ and $F_{.975,25,4}$ by $F_{.025,24,4} = 0.296$ and $F_{.975,24,4} = 8.51$. The

confidence interval is approximately $\left(\dfrac{s_X^2}{s_Y^2} F_{0.025,24,4}, \dfrac{s_X^2}{s_Y^2} F_{0.975,24,4} \right) =$

$\left(\dfrac{73.6}{86.9}(0.296), \dfrac{73.6}{86.9}(8.51) \right) = (0.251, 7.21)$. Because the confidence interval contains 1, it

supports the conclusion of Question 9.3.1 to accept H_0 that the variances are equal.

9.5.9 Since $\dfrac{S_Y^2/\sigma_Y^2}{S_X^2/\sigma_X^2}$ has an F distribution with $m-1$ and $n-1$ degrees of freedom,

$$P\left(F_{\alpha/2,m-1,n-1} \leq \frac{S_Y^2/\sigma_Y^2}{S_X^2/\sigma_X^2} \leq F_{1-\alpha/2,m-1,n-1} \right) = 1 - \alpha \text{ or}$$

$$P\left(\frac{S_X^2}{S_Y^2} F_{\alpha/2,m-1,n-1} \leq \frac{\sigma_X^2}{\sigma_Y^2} \leq \frac{S_X^2}{S_Y^2} F_{1-\alpha/2,m-1,n-1} \right) = 1 - \alpha$$

The inequality provides the confidence interval of Theorem 9.5.2.

9.5.11 The approximate normal distribution implies that

$$P\left(-z_\alpha \leq \frac{\dfrac{X}{n} - \dfrac{Y}{m} - (p_X - p_Y)}{\sqrt{\dfrac{(X/n)(1-X/n)}{n} + \dfrac{(Y/m)(1-Y/m)}{m}}} \leq z_\alpha \right) = 1 - \alpha$$

or $P\left(-z_\alpha \sqrt{\dfrac{(X/n)(1-X/n)}{n} + \dfrac{(Y/m)(1-Y/m)}{m}} \leq \dfrac{X}{n} - \dfrac{Y}{m} - (p_X - p_Y) \right.$

$\left. \leq z_\alpha \sqrt{\dfrac{(X/n)(1-X/n)}{n} + \dfrac{(Y/m)(1-Y/m)}{m}} \right) = 1 - \alpha$

which implies that

$P\left(-\left(\dfrac{X}{n} - \dfrac{Y}{m}\right) - z_\alpha \sqrt{\dfrac{(X/n)(1-X/n)}{n} + \dfrac{(Y/m)(1-Y/m)}{m}} \leq -(p_X - p_Y) \right.$

$\left. \leq -\left(\dfrac{X}{n} - \dfrac{Y}{m}\right) + z_\alpha \sqrt{\dfrac{(X/n)(1-X/n)}{n} + \dfrac{(Y/m)(1-Y/m)}{m}} \right) = 1 - \alpha$

Multiplying the inequality by -1 yields the confidence interval.

Chapter 10

Section 10.2

10.2.1 Let X_i = number of students with a score of i, $i = 1, 2, 3, 4, 5$. Then $P(X_1 = 0, X_2 = 0, X_3 = 1,$
$X_4 = 2, X_5 = 3) = \dfrac{6!}{0!0!1!2!3!}(0.116)^0(0.325)^0(0.236)^1(0.211)^2(0.112)^3 = 0.000886.$

10.2.3 Let Y denote a person's blood pressure and let X_1, X_2, and X_3 denote the number of individuals with blood pressures less than 140, between 140 and 160, and over 160, respectively. If $\mu = 124$ and $\sigma = 13.7$, $p_1 = P(Y < 140) = P\left(Z < \dfrac{140 - 124}{13.7}\right) = 0.8790,$

$p_2 = P(140 \leq Y \leq 160) = P\left(\dfrac{140 - 124}{13.7} \leq Z \leq \dfrac{160 - 124}{13.7}\right) = 0.1167,$ and $p_3 = 1 - p_1 - p_2 =$
$0.0043.$ Then $P(X_1 = 6, X_2 = 3, X_3 = 1) = \dfrac{10!}{6!3!1!}(0.8790)^6(0.1167)^3(0.0043)^1 = 0.00265.$

10.2.5 Let Y denote the distance between the pipeline and the point of impact. Let X_1 denote the number of missiles landing within 20 yards to the left of the pipeline, let X_2 denote the number of missiles landing within 20 yards to the right of the pipeline, and let X_3 denote the number of missiles for which $|y| > 20$. By the symmetry of $f_Y(y)$, $p_1 = P(-20 \leq Y \leq 0) = \dfrac{5}{18}$

$= P(0 \leq Y \leq 20) = p_2$ (so $p_3 = P(|Y| > 20) = 1 - \dfrac{5}{18} - \dfrac{5}{18} = \dfrac{8}{18}$). Therefore, $P(X_1 = 2, X_2 = 4,$

$X_3 = 0) = \dfrac{6!}{2!4!0!}\left(\dfrac{5}{18}\right)^2\left(\dfrac{5}{18}\right)^4\left(\dfrac{8}{18}\right)^0 = 0.00649.$

10.2.7 (a) $p_1 = P\left(0 \leq Y < \dfrac{1}{4}\right) = \int_0^{1/4} 3y^2 dy = \dfrac{1}{64}, p_2 = P\left(\dfrac{1}{4} \leq Y < \dfrac{1}{2}\right) = \int_{1/4}^{1/2} 3y^2 dy = \dfrac{7}{64},$

$p_3 = P\left(\dfrac{1}{2} \leq Y < \dfrac{3}{4}\right) = \int_{1/2}^{3/4} 3y^2 dy = \dfrac{19}{64},$ and $p_4 = P\left(\dfrac{3}{4} \leq Y \leq 1\right) = \int_{3/4}^1 3y^2 dy = \dfrac{37}{64}.$ Then

$f_{X_1, X_2, X_3, X_4}(3, 7, 15, 25) = P(X_1 = 3, X_2 = 7, X_3 = 15, X_4 = 25) =$

$\dfrac{50!}{3!7!15!25!}\left(\dfrac{1}{64}\right)^3\left(\dfrac{7}{64}\right)^7\left(\dfrac{19}{64}\right)^{15}\left(\dfrac{37}{64}\right)^{25}.$

(b) By Theorem 10.2.2, X_3 is a binomial random variable with parameters n (= 50) and

$p_3\left(=\dfrac{19}{64}\right).$ Therefore, $\text{Var}(X_3) = np_3(1 - p_3) = 50\left(\dfrac{19}{64}\right)\left(\dfrac{45}{64}\right) = 10.44.$

10.2.9 Assume that $M_{X_1, X_2, X_3}(t_1, t_2, t_3) = \left(p_1 e^{t_1} + p_2 e^{t_2} + p_3 e^{t_3}\right)^n.$ Then $M_{X_1, X_2, X_3}(t_1, 0, 0) =$

$E(e^{t_1 X_1}) = \left(p_1 e^{t_1} + p_2 + p_3\right)^n = (1 - p_1 + p_1 e^{t_1})^n$ is the mgf for X_1. But the latter has the form of the mgf for a binomial random variable with parameters n and p_1.

Section 10.3

10.3.1 $\sum_{i=1}^{t}\dfrac{(X_i - np_i)^2}{np_i} = \sum_{i=1}^{t}\dfrac{(X_i^2 - 2np_iX_i + n^2p_i^2)}{np_i} = \sum_{i=1}^{t}\dfrac{X_i^2}{np_i} - 2\sum_{i=1}^{t}X_i + n\sum_{i=1}^{t}p_i = \sum_{i=1}^{t}\dfrac{X_i^2}{np_i} - n.$

10.3.3 If the sampling is presumed to be <u>with</u> replacement, the number of white chips selected would follow a binomial distribution. Specifically, $\pi_1 = P(0 \text{ whites are drawn}) =$

$$\binom{2}{0}\left(\frac{4}{10}\right)^0\left(\frac{6}{10}\right)^2 = 0.36, \; \pi_2 = P(1 \text{ white is drawn}) = \binom{2}{1}\left(\frac{4}{10}\right)^1\left(\frac{6}{10}\right)^1 = 0.48, \text{ and}$$

$$\pi_3 = P(2 \text{ whites are drawn}) = \binom{2}{2}\left(\frac{4}{10}\right)^2\left(\frac{6}{10}\right)^0 = 0.16. \text{ The form of the } \alpha = 0.10 \text{ decision rule}$$

would be the same as in Question 10.3.2—reject H_0 if $d \geq \chi^2_{.90,2} = 4.605$. In this case, though,

$$d = \frac{(35 - 100(0.36))^2}{100(0.36)} + \frac{(55 - 100(0.48))^2}{100(0.48)} + \frac{(10 - 100(0.16))^2}{100(0.16)} = 3.30. \quad \text{Although the}$$

binomial fit is not as good as the hypergeometric fit done in Question 10.3.2, the null hypothesis that the sampling occurred with replacement is not rejected.

10.3.5 Let $p = P(\text{baby is born between midnight and 4 A.M.})$. Test H_0: $p = \dfrac{1}{6}$ versus H_1: $p \neq \dfrac{1}{6}$.

Let n = number of births = 2650 and X = number of babies born between midnight and 4 A.M. = 494. From Theorem 6.3.1, H_0 should be rejected if z is either ≤ -1.96 or

$\geq 1.96 \;(= \pm z_{.025})$. Here $z = \dfrac{494 - 2650(1/6)}{\sqrt{2650(1/6)(5/6)}} = 2.73$, so H_0 is rejected. These two test

procedures are equivalent: If one rejects H_0, so will the other. Notice that $z^2_{.025} = (1.96)^2 = 3.84 = \chi^2_{.95,1}$ and (except for a small rounding error) $z^2 = (2.73)^2 = 7.45 = \chi^2 = 7.44$.

10.3.7 Listed in the accompanying table are the observed and expected numbers of M&Ms of each color. Let p_1 = true proportion of browns, p_2 = true proportion of yellows, and so on.

Color	X_i	π_i	$E(X_i) = 1527 \cdot \pi_i$
Brown	455	0.30	458.1
Yellow	343	0.20	305.4
Red	318	0.20	305.4
Orange	152	0.10	152.7
Blue	130	0.10	152.7
Green	129	0.10	152.7
	1527	1.00	1527.0

To test H_0: $p_1 = 0.30, p_2 = 0.20, \ldots, p_6 = 0.10$ versus H_1: at least one $p_i \neq \pi_i$, reject H_0 if

$d \geq \chi^2_{.95,5} = 11.070$. But $d = \dfrac{(455 - 458.1)^2}{458.1} + \ldots + \dfrac{(129 - 152.7)^2}{152.7} = 12.23$, so H_0 is rejected

(these particular observed frequencies are not consistent with the company's intended probabilities).

10.3.9 Let the random variable X denote the length of a World Series. Then $P(X=4) = \pi_1 = P(\text{AL}$ wins in 4) + $P(\text{NL wins in 4}) = 2 \cdot P(\text{AL wins in 4}) = 2\left(\dfrac{1}{2}\right)^4 = \dfrac{1}{8}$. Similarly, $P(X=5) = \pi_2 = 2 \cdot P(\text{AL wins in 5}) = 2 \cdot P(\text{AL wins exactly 3 of first 4 games}) \cdot P(\text{AL wins 5th game}) = 2 \cdot \binom{4}{3}\left(\dfrac{1}{2}\right)^3\left(\dfrac{1}{2}\right)^1 \cdot \dfrac{1}{2} = \dfrac{1}{4}$. Also, $P(X=6) = \pi_3 = 2 \cdot P(\text{AL wins exactly 3 of first 5 games}) \cdot$

$P(\text{AL wins 6th game}) = 2 \cdot \binom{5}{3}\left(\dfrac{1}{2}\right)^3\left(\dfrac{1}{2}\right)^2 \cdot \left(\dfrac{1}{2}\right) = \dfrac{5}{16}$, and $P(X=7) = \pi_4 = 1 - P(X=4) -$

$P(X=5) - P(X=6) = \dfrac{5}{16}$. Listed in the table is the information necessary for calculating the goodness-of-fit statistic d. The "Bernoulli model" is rejected if $d \geq \chi^2_{.90,3} = 6.251$. For these

data, $d = \dfrac{(9-6.25)^2}{6.25} + \dfrac{(11-12.50)^2}{12.50} + \dfrac{(8-15.625)^2}{15.625} + \dfrac{(22-15.625)^2}{15.625} = 7.71$, so H_0 is rejected.

Number of games	Number of years	$50 \cdot \pi_i$
4	9	6.25
5	11	12.50
6	8	15.625
7	22	15.625
	50	50.000

10.3.11 Listed is the frequency distribution for the 70 y_i's using classes of width 10 starting at 220. If normality holds, each π_i is an integral of the normal pdf having $\mu = 266$ and $\sigma = 16$.

Duration, y	Freq.	π_i	$E(X_i)$
$220 \leq y < 230$	1	0.0122	0.854 ⎫
$230 \leq y < 240$	5	0.0394	2.758 ⎬ 11.109
$240 \leq y < 250$	10	0.1071	7.497 ⎭
$250 \leq y < 260$	16	0.1933	13.531
$260 \leq y < 270$	23	0.2467	17.269
$270 \leq y < 280$	7	0.2119	14.833
$280 \leq y < 290$	6	0.1226	8.582 ⎫
$290 \leq y < 300$	2	0.0668	4.676 ⎬ 13.258
	70	1.0000	70.000

For example, $\pi_2 = P(230 \le Y < 240) = P\left(\dfrac{230-266}{16} \le \dfrac{Y-266}{16} < \dfrac{240-266}{16}\right) =$

$P(-2.25 \le Z < -1.63) = 0.0394.$ (Note: To account for all the area under $f_Y(y)$, the intervals defining the first and last classes need to be extended to $-\infty$ and $+\infty$, respectively. That is, $\pi_1 = P(-\infty < Y < 230)$ and $\pi_8 = P(290 \le Y < \infty)$.) Some of the expected frequencies ($= 70 \cdot \pi_i$) are too small (i.e., less than 5) for the χ^2 approximation to be fully adequate, so the first three classes need to be combined, as do the last two. With $t = 5$ final classes, then, the normality assumption is rejected if $d \ge \chi^2_{.90,4} = 7.779$. Here,

$d = \dfrac{(16-11.109)^2}{11.109} + ... + \dfrac{(8-13.258)^2}{13.258} = 10.73$, so we would reject the null hypothesis that pregnancy durations are normally distributed.

Section 10.4

10.4.1 Let $p = P$(voter says "yes"). Then $\hat{p} = \dfrac{\text{number of yeses}}{\text{number of voters}} = \dfrac{30(0) + 56(1) + 73(2) + 41(3)}{600} =$

0.54, so the H_0 model to be tested is $P(i \text{ yeses}) = \dbinom{3}{i}(0.54)^i(0.46)^{3-i}$, $i = 0, 1, 2, 3$. Detailed

in the accompanying table are the relevant observed and expected frequencies. At the $\alpha = 0.05$ level, the binomial model should be rejected if $d_1 \ge \chi^2_{.95,4-1-1} = 5.991$.

No. saying "yes"	Freq.	$\hat{p}_i$	$200 \cdot \hat{p}_i$
0	30	0.097	19.4
1	56	0.343	68.6
2	73	0.402	80.4
3	41	0.157	31.4
	200	1.000	200.0

But $d_1 = \dfrac{(30-19.4)^2}{19.4} + \dfrac{(56-68.6)^2}{68.6} + \dfrac{(73-80.4)^2}{80.4} + \dfrac{(41-31.4)^2}{31.4} = 11.72$, implying that the binomial model is inadequate in this particular context (probably because the trials are not likely to be independent, which is one of the model's assumptions).

10.4.3 Here the H_0 model is $P(y \text{ infected plants}) = e^{-\lambda}(\hat{\lambda})^i / i!$, $i = 0, 1, 2, \ldots$, where

$\hat{\lambda} = \dfrac{38(0) + 57(1) + \ldots + 1(12)}{270} = 2.53$. As the table clearly shows, the Poisson model is inappropriate for these data. The disagreements between the observed and expected frequencies are considerable—$d_1 = \dfrac{(38 - 21.52)^2}{21.52} + \ldots + \dfrac{(28 - 11.88)^2}{11.88} = 46.75$, which greatly exceeds the $\alpha = 0.05$ critical value, $\chi^2_{.95, 7-1-1} = 11.070$. The independence assumption would not hold if the infestation was contagious (which is likely to be the case).

No. of Infected Plants	No. of Quadrats	$\hat{p}_i$	$270 \cdot \hat{p}_i$
0	38	0.0797	21.52
1	57	0.2015	54.41
2	68	0.2549	68.82
3	47	0.2150	58.05
4	23	0.1360	36.72
5	9	0.0688	18.58
6	10	0.0290	7.83
7	7	0.0105	2.84
8	3	0.0033	0.89
9	4	0.0009	0.24
10	2	0.0002	0.05
11	1	0.0001	0.03
12	1	0.0000	0.00
13+	0	0.0000	0.00
	270	1.0000	270.00

(6 through 13+ bracketed: 11.88)

10.4.5 Under H_0, the intervals between shutdowns should be described by an exponential pdf, $f_Y(y) = \hat{\lambda} e^{-\hat{\lambda} y}$, $y > 0$, where $\hat{\lambda} = 1/\overline{y}$ (recall Theorem 4.2.3). Here, the sample mean can be approximated by assigning each observation in a range a value equal to the midpoint of that range. Therefore, $\overline{y} \doteq \dfrac{130(0.5) + 41(1.5) + \ldots + 1(7.5)}{211} = 1.22$, which makes $\hat{\lambda} = 0.82$.

Moreover, each $\hat{p}_i$ is an area under $f_Y(y)$. For example, $\hat{p}_1 = \int_0^1 0.82 e^{-0.82y} dy = 0.56$. The complete set of $\hat{p}_i$'s and estimated expected frequencies are listed in the accompanying table. Using $t = 5$ final classes, we should reject the exponential model if $d_1 \geq \chi^2_{.95, 5-1-1} = 7.815$. But $d_1 = \dfrac{(130 - 118.16)^2}{118.16} + \ldots + \dfrac{(7 - 8.01)^2}{8.01} = 4.2$, so H_0 is not rejected.

Interval	Freq.	$\hat{p}_i$	$211 \cdot \hat{p}_i$
$0 \le y < 1$	130	0.560	118.16
$1 \le y < 2$	41	0.246	51.91
$2 \le y < 3$	25	0.109	23.00
$3 \le y < 4$	8	0.047	9.92
$4 \le y < 5$	2	0.021	4.43
$5 \le y \le 6$	3	0.010	2.11
$6 \le y < 7$	1	0.004	0.84
$y \ge 7$	1	0.003	0.63
	211	1.000	211.00

(braces grouping last four expected values: 8.01)

10.4.7 If $p = P$(child is a boy), $\hat{p} = \dfrac{\text{number of boys}}{\text{number of children}} = \dfrac{24(0) + 64(1) + 32(2)}{240} = 0.533$, so the

hypotheses to be tested are H_0: $P(i \text{ boys}) = \dbinom{2}{i}(0.533)^i(0.467)^{2-i}$, $i = 0, 1, 2$, versus

H_1: $P(i \text{ boys}) \ne \dbinom{2}{i}(0.533)^i(0.467)^{2-i}$, $i = 0, 1, 2$. Summarized in the table are the observed

and expected numbers of families with 0, 1, and 2 boys. Given that t = number of classes = 3 and that 1 parameter has been estimated, H_0 should be rejected if $d_1 \ge \chi^2_{.95,3-1-1} = 3.841$. But

$d_1 = \dfrac{(24-26.2)^2}{26.2} + \dfrac{(64-59.7)^2}{59.7} + \dfrac{(32-34.1)^2}{34.1} = 0.62$, implying that the binomial model

should not be rejected.

No. of boys	Freq.	$\hat{p}_i$	$120 \cdot \hat{p}_i$
0	24	0.2181	26.2
1	64	0.4978	59.7
2	32	0.2841	34.1
	120	1.0000	120.0

10.4.9 Given that $\hat{\lambda} = 3.87$, the model to fit under H_0 is $p_X(i) = e^{-3.87}(3.87)^i/i!$, $i = 0, 1, 2, ...$ Multiplying the latter probabilities by 2608 gives the complete set of estimated expected frequencies, as shown in the table. No classes need to be combined, so $t = 12$ and one parameter has been estimated. Let $\alpha = 0.05$. Then H_0 should be rejected if $d_1 \geq \chi^2_{.95,12-1-1} = 18.307$. But $d_1 = \dfrac{(57-54.51)^2}{54.51} + \dfrac{(203-210.47)^2}{210.47} + ... + \dfrac{(6-6.00)^2}{6.00} = 12.92$, implying that the Poisson model should not be rejected.

No. detected, i	Freq.	$\hat{p}_i \, (= p_X(k))$	$2608 \cdot \hat{p}_i$
0	57	0.0209	54.51
1	203	0.0807	210.47
2	383	0.1562	407.37
3	525	0.2015	525.51
4	532	0.1949	508.30
5	408	0.1509	393.55
6	273	0.0973	253.76
7	139	0.0538	140.31
8	45	0.0260	67.81
9	27	0.0112	29.21
10	10	0.0043	11.21
11+	6	0.0023	6.00
	2608	1.0000	2608.00

10.4.11 The MLE for p is the reciprocal of the sample mean. Here, $\hat{p} = \dfrac{50}{4(1) + 13(2) + ... + 1(9)} = 0.26$, so the H_0 model becomes $p_X(i) = (1 - 0.26)^{i-1}(0.26)$, $i = 1, 2, ...$ Combining the last five classes (see the accompanying table) makes $t = 5$. Let $\alpha = 0.05$. Then H_0 should be rejected if $d_1 \geq \chi^2_{.95,5-1-1} = 7.815$. In this case, $d_1 = \dfrac{(4-13.00)^2}{13.00} + \dfrac{(13-9.62)^2}{9.62} + ... + \dfrac{(16-15.00)^2}{15.00} = 9.23$, which suggests that the 50 observations did not come from a geometric pdf.

Outcome	Freq.	$\hat{p}_i \, (= p_X(k))$	$50 \cdot \hat{p}_i$	
1	4	0.2600	13.00	
2	13	0.1924	9.62	
3	10	0.1424	7.12	
4	7	0.1054	5.27	
5	5	0.0780	3.90	
6	4	0.0577	2.89	
7	3	0.0427	2.14	15.00
8	3	0.0316	1.58	
9+	1	0.0898	4.49	
	50	1.0000	50.00	

Section 10.5

10.5.1 To test H_0: Telephone listing and home ownership are independent at the $\alpha = 0.05$ level, reject H_0 if $d_2 \geq \chi^2_{.95,(2-1)(2-1)} = 3.841$. If α is increased to 0.10, the critical value reduces to $\chi^2_{.90,1} = 2.706$. Based on the expected frequencies predicted by H_0 (see the accompanying table), $d_2 = 2.77$ so H_0 is rejected at the $\alpha = 0.10$ level, but not at the $\alpha = 0.05$ level.

	Listed	Unlisted	
Own	628	146	774
	(619.20)	(154.80)	
Rent	172	54	226
	(180.80)	(45.20)	
	800	200	1000

10.5.3 To test H_0: Delinquency and birth order are independent versus H_1: Delinquency and birth order are dependent at the $\alpha = 0.01$ level, reject the null hypothesis if $d_2 \geq \chi^2_{.99,(4-1)(2-1)} =$ 11.345. Here, $d_2 = \dfrac{(24-45.59)^2}{45.59} + ... + \dfrac{(70-84.05)^2}{84.05} = 42.25$, suggesting that delinquency and birth order <u>are</u> related.

	Delinquent	Not delinquent	
Oldest	24	450	474
	(45.59)	(428.41)	
In between	29	312	341
	(32.80)	(308.20)	
Youngest	35	211	246
	(23.66)	(222.34)	
Only	23	70	93
	(8.95)	(84.05)	
	111	1043	1154

10.5.5 The null hypothesis that adults' self-perception and attitude toward small cars are independent should be rejected at the $\alpha = 0.01$ level if $d_2 \geq \chi^2_{.99,(3-1)(3-1)} = 13.277$. These data suggest that the two are <u>not</u> independent (and H_0 should be rejected)—
$d_2 = \dfrac{(79-61.59)^2}{61.59} + ... + \dfrac{(42-28.76)^2}{28.76} = 27.29.$

	Cautious	Middle	Confident	
Favorable	79	58	49	186
	(61.59)	(62.21)	(62.21)	
Neutral	10	8	9	27
	(8.94)	(9.03)	(9.03)	
Unfavorable	10	34	42	86
	(28.47)	(28.76)	(28.76)	
	99	100	100	299

10.5.7 Given that $\alpha = 0.05$, the null hypothesis that early upbringing and aggressiveness later in life are independent is rejected if $d_2 \geq \chi^2_{.95,(2-1)(2-1)} = 3.841$. But $d_2 = \dfrac{(27-40.25)^2}{40.25} + \ldots + \dfrac{(93-106.25)^2}{106.25} = 12.61$, so H_0 <u>is</u> rejected—mice raised by foster mothers appear to be more aggressive than mice raised by their natural mothers.

	Natural mother	Foster mother	
No. fighting	27	47	74
	(40.25)	(33.75)	
No. not fighting	140	93	233
	(126.75)	(106.25)	
	167	140	307

10.5.9 Let $\alpha = 0.05$. To test the null hypothesis that annual return and portfolio turnover are independent, we should reject H_0 if $d_2 \geq \chi^2_{.95,(2-1)(2-1)} = 3.841$. Based on the table below, the value of d_2 in this case is $2.20 \left(= \dfrac{(11-13.86)^2}{13.86} + \ldots + \dfrac{(24-26.86)^2}{26.86} \right)$, so the appropriate conclusion would be that annual return and portfolio turnover <u>are</u> independent.

		Annual return		
		$\leq 10\%$	$> 10\%$	
Portfolio	$\geq 100\%$	11	10	21
return		(13.86)	(7.14)	
	$<100\%$	55	24	79
		(52.14)	(26.86)	
		66	34	100

Chapter 11

Section 11.2

11.2.1 $b = \dfrac{n\sum_{i=1}^{n} x_i y_i - \left(\sum_{i=1}^{n} x_i\right)\left(\sum_{i=1}^{n} y_i\right)}{n\left(\sum_{i=1}^{n} x_i^2\right) - \left(\sum_{i=1}^{n} x_i\right)^2} = \dfrac{15(20{,}127.47) - (249.8)(1{,}200.6)}{15(4200.56) - (249.8)^2} = 3.291$

$a = \dfrac{\sum_{i=1}^{n} y_i - b\sum_{i=1}^{n} x_i}{n} = \dfrac{1{,}200.6 - 3.291(249.8)}{15} = 25.234$

Then $y = 25.234 + 3.291x$; $y(18) = 84.5°F$

11.2.3 $b = \dfrac{n\sum_{i=1}^{n} x_i y_i - \left(\sum_{i=1}^{n} x_i\right)\left(\sum_{i=1}^{n} y_i\right)}{n\left(\sum_{i=1}^{n} x_i^2\right) - \left(\sum_{i=1}^{n} x_i\right)^2} = \dfrac{9(24{,}628.6) - (234)(811.3)}{9(10{,}144) - (234)^2} = 0.8706$

$a = \dfrac{\sum_{i=1}^{n} y_i - b\sum_{i=1}^{n} x_i}{n} = \dfrac{811.3 - 0.8706(234)}{9} = 67.5088$

As an example of calculating a residual, consider $x_2 = 4$. Then the corresponding residual is $y_2 - \hat{y}_2 = 71.0 - [67.5088 + 0.8706(4)] = 0.0098$. The complete set of residuals, rounded to two decimal places is

x_i	$y_i - \hat{y}_i$
0	−0.81
4	0.01
10	0.09
15	0.03
21	−0.09
29	0.14
36	0.55
51	1.69
68	−1.61

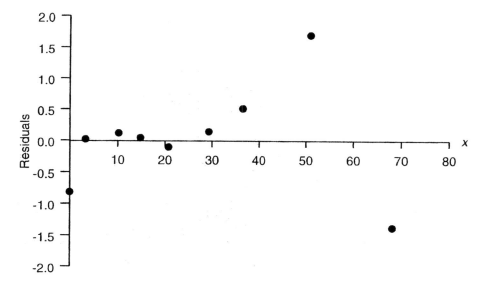

A straight line appears to fit these data.

11.2.5 The value 12 is too "far" from the data observed.

11.2.7 $b = \dfrac{13(48,593,986) - (54,975)(11,431)}{13(237,083,328) - (54,975)^2} = 0.0552$

$a = \dfrac{11,431 - 0.0552(54,975)}{13} = 645.88$

The least squares line is $645.88 + 0.0552x$. The plot of the data and least square line is:

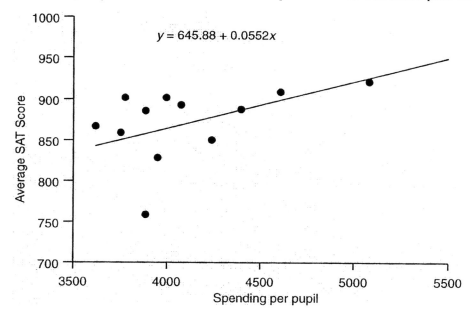

The linear fit for x values less than \$4300 is not very good, suggesting a search for other contributing variables in the x range of \$3500 to \$4200.

11.2.9 $b = \dfrac{9(7,439.37) - (41.56)(1,416.1)}{9(289.4222) - (41.56)^2} = 9.23$

$a = \dfrac{(1,416.1) - 9.23(41.56)}{9} = 114.72$

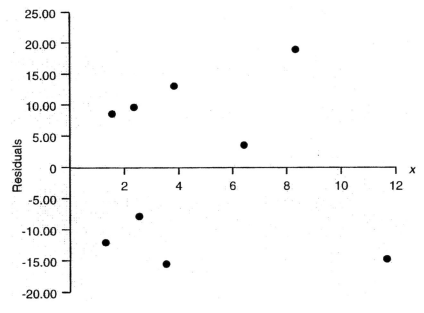

A linear relationship seems reasonable.

11.2.11 $b = \dfrac{11(1141) - (111)(100)}{11(1277) - (111)^2} = 0.84$

$a = \dfrac{1072 - 0.84(111)}{11} = 0.61$

The least squares line is $y = 0.61 + 0.84x$. The residuals given in the table below are large relative to the x values, which suggests that the linear fit is inadequate.

x_i	$y_i - \hat{y}_i$
7	−3.5
13	−1.5
14	−1.4
6	−0.7
14	2.6
15	1.8
4	3.0
8	2.7
7	−2.5
9	0.8
14	−1.4

11.2.13 When $\bar{x}$ is substituted for x in the least-squares line equation, we obtain
$$y = a + b\bar{x} = \bar{y} - b\bar{x} + b\bar{x} = \bar{y}$$

11.2.15 For these data $\sum_{i=1}^{n} d_i v_i = 95,161.2$, and $\sum_{i=1}^{n} d_i^2 = 2,685,141$.

Then $H = \dfrac{\sum_{i=1}^{n} d_i v_i}{\sum_{i=1}^{n} d_i^2} = \dfrac{95,161.2}{2,685,141} = 0.03544$.

11.2.17 $b = \dfrac{\sum_{i=1}^{n} x_i y_i - a \cdot \sum_{i=1}^{n} x_i}{\sum_{i=1}^{n} x_i^2} = \dfrac{1513 - 100(45)}{575.5} = -5.19$, so $y = 100 - 5.19x$.

11.2.19 (a)   The sums needed are $\sum_{i=1}^{13} x_i^2 = 54,537$, $\sum_{i=1}^{13} x_i y_i = 3,329.4$

Then $b = \dfrac{3329.4}{54,437} = 0.0612$

(b) $y(120) = 0.612(\$120) = \$7.34$ million

11.2.21 To fit the model $y = ae^{bx}$, we note that $\ln y$ is linear with x. Then

$$b = \dfrac{n\sum_{i=1}^{n} x_i \ln y_i - \left(\sum_{i=1}^{n} x_i\right)\left(\sum_{i=1}^{n} \ln y_i\right)}{n\left(\sum_{i=1}^{n} x_i^2\right) - \left(\sum_{i=1}^{n} x_i\right)^2}$$

$$= \dfrac{11(51.6005) - (66)(6.19815)}{11(506) - 66^2} = 0.131$$

$$\ln a = \dfrac{\sum_{i=1}^{n} \ln y_i - b\sum_{i=1}^{n} x_i}{n} = \dfrac{6.19815 - (0.131)(66)}{11} = -0.2225$$

Then $a = e^{-0.2225} = 0.8005$, and the model is $y = 0.8005e^{0.131x}$.

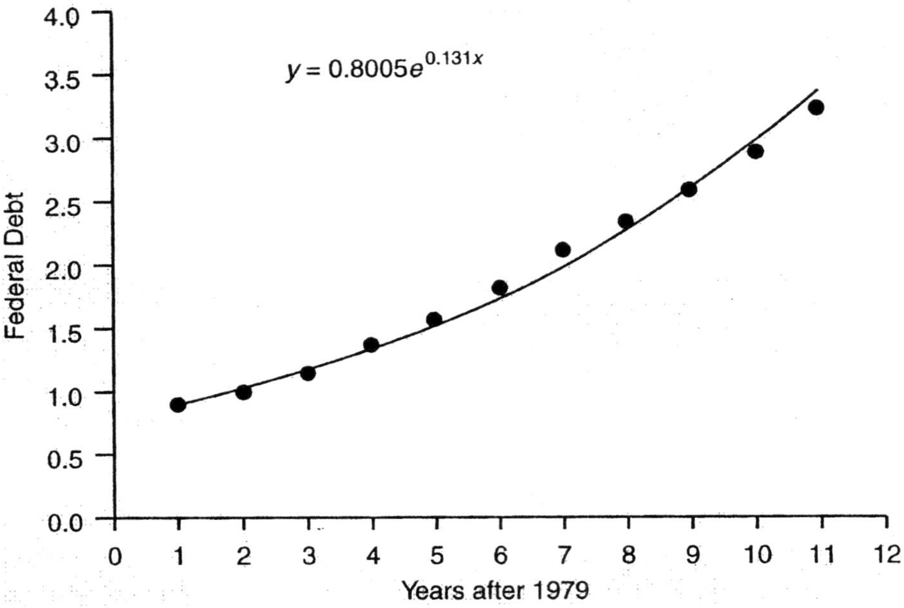

$y = 0.8005e^{0.131x}$
Federal Debt
4.0
3.5
3.0
2.5
2.0
1.5
1.0
0.5
0.0
0 1 2 3 4 5 6 7 8 9 10 11 12
Years after 1979

11.2.23 (b) $b = \dfrac{10(133.68654) - (55)(22.78325)}{10(385) - 55^2} = 0.10156$

$\ln a = \dfrac{22.78325 - (0.10156)(55)}{10} = 1.71975$

Then $a = e^{1.719745} = 5.5831$, and the model is $y = 5.5831e^{-0.10156x}$

(c)

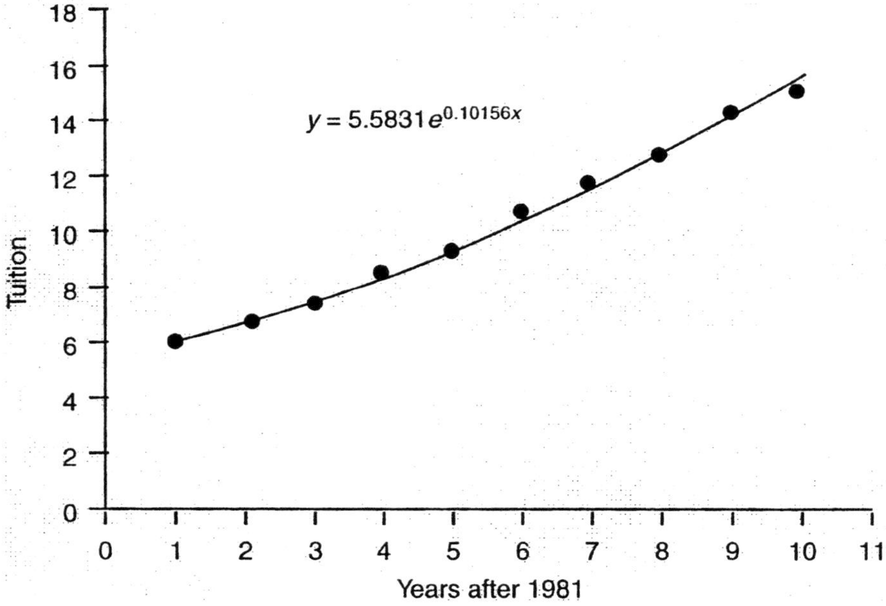

$y = 5.5831e^{0.10156x}$
Tuition
18
16
14
12
10
8
6
4
2
0
0 1 2 3 4 5 6 7 8 9 10 11
Years after 1981

(d) $y(37) + y(38) + y(39) + y(40) = 239.2 + 264.8 + 293.1 + 324.5 = 1{,}121.6$, or \$1,121,600.

11.2.25 $b = \dfrac{7(0.923141) - (-0.067772)(7.195129)}{7(0.0948679) - (-0.067772)^2} = 10.538;$

$\log a = \dfrac{1}{7}(7.195129) - \dfrac{10.538}{7}(-0.067772) = 1.1299$

Then $a = 10^{1.1299} = 13.487$. The model is $13.487x^{10.538}$.

The table below gives a comparison of the model values and the observed y_i's.

x_i	y_i	Model
0.98	25.000	10.901
0.74	0.950	0.565
1.12	200.000	44.522
1.34	150.000	294.677
0.87	0.940	3.109
0.65	0.090	0.144
1.39	260.000	433.516

11.2.27 $b = \dfrac{4(36.95941) - (11.55733)(12.08699)}{4(34.80999) - 11.55733^2} = 1.43687$

$\log a = \dfrac{12.08699 - 1.43687(11.55733)}{4} = -1.12985$

$a = 10^{-1.12985} = 0.07416$. The model is $y = 0.07416x^{1.43687}$

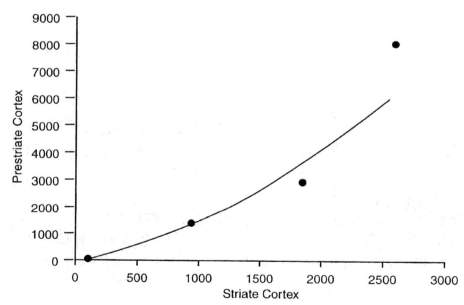

11.2.29 (d)   If $y = \dfrac{1}{a+bx}$, then $\dfrac{1}{y} = a + bx$ and $1/y$ is linear with x

(e)   If $y = \dfrac{x}{a+bx}$, then $\dfrac{1}{y} = \dfrac{a+bx}{x} = b + a\dfrac{1}{x}$, and $1/y$ is linear with $1/x$.

(f)   If $y = 1 - e^{-x^b/a}$, then $1 - y = e^{-x^b/a}$, and $\dfrac{1}{1-y} = e^{x^b/a}$. Taking the ln of both sides

gives $\ln\dfrac{1}{1-y} = x^b/a$. Taking the ln again yields

$\ln\ \ln\dfrac{1}{1-y} = -\ln a + b \ln x$, and $\ln\ \ln\dfrac{1}{1-y}$ is linear with $\ln x$.

11.2.31 Let $y' = \ln\left(\dfrac{60-y}{y}\right)$. We find the linear relationship between x and y'. The needed sums are

$$\sum_{i=1}^{8} x_i = 352, \quad \sum_{i=1}^{8} x_i^2 = 16160, \quad \sum_{i=1}^{8} y_i' = -2.39572, \quad \sum_{i=1}^{n} x_i y_i' = -194.88216$$

$$b = \frac{8(-194.88216) - 352(-2.39572)}{8(16160) - 352^2} = -0.13314$$

$$a = \frac{-2.39572 - (-0.13314)(352)}{8} = 5.55870$$

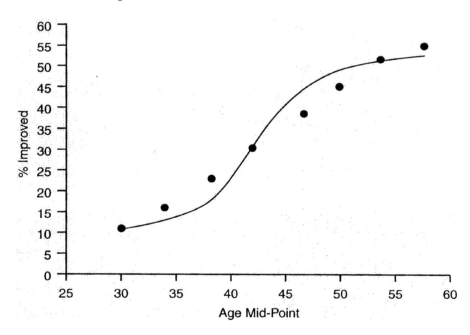

Chapter 11

Section 11.3

11.3.1 $\beta_1 = \dfrac{4(93) - 10(40.2)}{4(30) - 10^2} = -1.5$

$\beta_0 = \dfrac{(40.2) - (-1.5)(10)}{4} = 13.8$

Thus, $y = 13.8 - 1.5x$. $t = \dfrac{\hat{\beta}_1 - \beta_1^0}{s / \sqrt{\sum\limits_{i=1}^{4}(x_i - \bar{x})^2}} = \dfrac{-1.5 - 0}{2.114/\sqrt{5}} = -1.59$

Since $-t_{.025,2} = -4.3027 < t = -1.59 < 4.3027 = t_{.025,2}$, accept H_0.

11.3.3 $t = \dfrac{\hat{\beta}_1 - \beta_1^0}{s / \sqrt{\sum\limits_{i=1}^{15}(x_i - \bar{x})^2}} = \dfrac{3.291 - 0}{3.829/\sqrt{40.55733}} = 5.47.$

Since $t = 5.47 > t_{0.005,13} = 3.0123$, reject H_0.

11.3.5 $\mathrm{Var}(\hat{\beta}_1) = \sigma^2 / \sum\limits_{i=1}^{9}(x_i - \bar{x})^2 = 45/60 = 0.75.$ The standard deviation of $\hat{\beta}_1 = \sqrt{0.75} = 0.866.$

$P\left(\left|\hat{\beta}_1 - \beta_1\right| < 1.5\right) = P\left(\dfrac{\left|\hat{\beta}_1 - \beta_1\right|}{0.866} < \dfrac{1.5}{0.866}\right) = P(|Z| < 1.73)$ for the standard normal random

variable Z. $P(Z > 1.73) = 1 - 0.9582 = 0.0418$, so $P(|Z| < 1.73) = 1 - 2(0.0418) = 0.9164$

11.3.7 The radius of the confidence interval is

$t_{.05,7} \dfrac{s\sqrt{\sum\limits_{i=1}^{n}x_i^2}}{\sqrt{9}\sqrt{\sum\limits_{i=1}^{n}(x_i - \bar{x})^2}} = 1.8946 \dfrac{(0.959)\sqrt{10144}}{\sqrt{9}\sqrt{4060}} = 0.957$

The center of the interval is $\hat{\beta}_0 = 67.508$. The interval $= (66.551, 68.465)$

11.3.9 $t = \dfrac{\hat{\beta}_1 - \beta_1^0}{s / \sqrt{\sum\limits_{i=1}^{11}(x_i - \bar{x})^2}} = \dfrac{0.84 - 0}{2.404/\sqrt{156.909}} = 4.38$

Since $t = 4.38 > t_{.025,9} = 2.2622$, reject H_0.

11.3.11 By Theorem 11.3.2, $E(\hat{\beta}_0) = \beta_0$, and $\mathrm{Var}(\hat{\beta}_0) = \dfrac{\sigma^2 \sum\limits_{i=1}^{n} x_i}{n \sum\limits_{i=1}^{n} (x_i - \bar{x})^2}$.

Now, $(\hat{\beta}_0 - \beta_0)/\sqrt{\mathrm{Var}(\hat{\beta}_0)}$ is normal, so $P\left(-z_{\alpha/2} < (\hat{\beta}_0 - \beta_0)/\sqrt{\mathrm{Var}(\hat{\beta}_0)} < z_{\alpha/2}\right) = 1 - \alpha.$

Then the confidence interval is

$(\hat{\beta}_0 - z_{\alpha/2}\sqrt{\mathrm{Var}(\hat{\beta}_0)},\ \hat{\beta}_0 + z_{\alpha/2}\sqrt{\mathrm{Var}(\hat{\beta}_0)}$ or

$$\left(\hat{\beta}_0 - z_{\alpha/2} \frac{\sigma\sqrt{\sum\limits_{i=1}^{n} x_i}}{\sqrt{n}\sqrt{\sum\limits_{i=1}^{n}(x_i - \bar{x})^2}},\ \hat{\beta}_0 + z_{\alpha/2} \frac{\sigma\sqrt{\sum\limits_{i=1}^{n} x_i}}{\sqrt{n}\sqrt{\sum\limits_{i=1}^{n}(x_i - \bar{x})^2}} \right)$$

11.3.13 Reject the null hypothesis if the statistic is $< \chi^2_{\alpha/2,n-2} = \chi^2_{.025,22} = 10.982$ or $> \chi^2_{1-\alpha/2,n-2} = \chi^2_{.975,22}$

$= 36.781$. The observed chi square is $\dfrac{(n-2)s^2}{\sigma_0^2} = \dfrac{(24-2)(18.2)}{12.6} = 31.778$, so do not reject H_0.

11.3.15 The value of is given to be 2.31, so $s^2 = 5.3361$. Then the confidence interval for σ^2 is

$$\left(\frac{(n-2)s^2}{\chi^2_{1-\alpha/2,n-2}}, \frac{(n-2)s^2}{\chi^2_{\alpha/2,n-2}} \right) = \left(\frac{(7)(5.3361)}{\chi^2_{.95,7}}, \frac{(7)(5.3361)}{\chi^2_{.05,7}} \right) = \left(\frac{37.3527}{14.067}, \frac{37.3527}{2.167} \right)$$

$= (2.655, 17.237)$

11.3.17 The radius of the 95% confidence interval is

$$2.0687(0.0113)\sqrt{\frac{1}{25} + \frac{(2.750 - 2.643)^2}{0.0367}} = 0.0139$$

The center is $\hat{y} = \hat{\beta}_0 + \hat{\beta}_1 x = 0.308 + 0.642(2.750) = 2.0735$. The confidence interval is $(2.0596, 2.0874)$

11.3.19 The radius of the 95% prediction interval is $2.7764(6.987)\sqrt{1 + \dfrac{1}{6} + \dfrac{(7-3.5)^2}{17.5}} = 26.504.$

The center is $\hat{y} = \hat{\beta}_0 + \hat{\beta}_1 x = 97.867 - 11.057(7) = 20.468.$

The 95% prediction interval for 1996 ($x = 7$) is $(-6.04, 46.97)$ so the worst-case scenario is 47 near-collisions.

11.3.21 The test statistic is $t = \dfrac{\hat{\beta}_1 - \hat{\beta}_1^*}{s\sqrt{\dfrac{1}{\sum_{i=1}^{6}(x_i - \bar{x})^2} + \dfrac{1}{\sum_{i=1}^{8}(x_i^* - \bar{x}^*)^2}}}$, where

$$s = \sqrt{\frac{5.983 + 13.804}{6 + 8 - 4}} = 1.407. \text{ Then } t = \frac{0.606 - 1.07}{1.407\sqrt{\dfrac{1}{31.33} + \dfrac{1}{46}}} = -1.42$$

Since the observed ratio is not less than $-t_{.05,10} = -1.8125$ the difference in slopes can be ascribed to chance. These data do not support further investigation.

11.3.23 The form given in the text is

$$\text{Var}(\hat{Y}) = \sigma^2 \left[\frac{1}{n} + \frac{(x - \bar{x})^2}{\sum_{i=1}^{n}(x_i - \bar{x})^2} \right].$$ Putting the sum in the brackets over a least common

denominator gives $\dfrac{1}{n} + \dfrac{(x - \bar{x})^2}{\sum_{i=1}^{n}(x_i - \bar{x})^2} = \dfrac{\sum_{i=1}^{n}(x_i - \bar{x})^2 + n(x - \bar{x})^2}{n\sum_{i=1}^{n}(x_i - \bar{x})^2}$

$$= \frac{\sum_{i=1}^{n} x_i^2 - n\bar{x}^2 + n(x^2 + \bar{x}^2 - 2x\bar{x})}{n\sum_{i=1}^{n}(x_i - \bar{x})^2} = \frac{\sum_{i=1}^{n} x_i^2 + nx^2 - 2nx\bar{x}}{n\sum_{i=1}^{n}(x_i - \bar{x})^2} = \frac{\sum_{i=1}^{n} x_i^2 + nx^2 - 2x\sum_{i=1}^{n} x_i}{n\sum_{i=1}^{n}(x_i - \bar{x})^2}$$

$$= \frac{\sum_{i=1}^{n}(x_i - x)^2}{n\sum_{i=1}^{n}(x_i - \bar{x})^2}. \text{ Thus } \text{Var}(\hat{Y}) = \frac{\sigma^2 \sum_{i=1}^{n}(x_i - x)^2}{n\sum_{i=1}^{n}(x_i - \bar{x})^2}.$$

Section 11.4

11.4.1

x	y	$f_{X,Y}$	xy	$xyf_{X,Y}$
1	1	1/36	1	1/36
1	2	1/36	2	2/36
1	3	1/36	3	3/36
1	4	1/36	4	4/36
1	5	1/36	5	5/36
1	6	1/36	6	6/36
2	2	2/36	4	8/36
2	3	1/36	6	6/36
2	4	1/36	8	8/36
2	5	1/36	10	10/36
2	6	1/36	12	12/36
3	3	3/36	9	27/36
3	4	1/36	12	12/36
3	5	1/36	15	15/36
3	6	1/36	18	18/36
4	4	4/36	16	64/36
4	5	1/36	20	20/36
4	6	1/36	24	24/36
5	5	5/36	25	125/36
5	6	1/36	30	30/36
6	6	6/36	36	216/36

$E(XY)$ is the sum of the last column $= \dfrac{616}{36}$

Clearly $E(X) = 7/2$.

$$E(Y) = 1\frac{1}{36} + 2\frac{2}{36} + 3\frac{5}{36} + 4\frac{7}{36} + 5\frac{9}{36} + 6\frac{11}{36} = \frac{161}{36}$$

$$\text{Cov}(X,Y) = E(XY) - E(X)E(Y) = \frac{616}{36} - \frac{7}{2}\cdot\frac{161}{36} = \frac{105}{72}$$

11.4.3 $\displaystyle\int_0^{2\pi} \cos x\,dx = \int_0^{2\pi} \sin x\,dx = \int_0^{2\pi} (\cos x)(\sin x)\,dx = 0$, so $E(X) = E(Y) = E(XY) = 0$.

Then $\text{Cov}(X, Y) = 0$. But X and Y are functionally dependent, $Y = \sqrt{1 - X^2}$, so they are probabilistically dependent.

11.4.5 $E(XY) = 1\dfrac{1+2(1)}{22} + 2\dfrac{2+2(1)}{22} + 3\dfrac{1+2(3)}{22} + 6\dfrac{2+2(3)}{22} = 80/22 = 40/11$

$E(X) = 1\dfrac{10}{22} + 2\dfrac{12}{22} = 34/22 = 17/11$

$E(X^2) = 1\dfrac{10}{22} + 4\dfrac{12}{22} = 58/22 = 29/11$

$E(Y) = 1\dfrac{7}{22} + 3\dfrac{15}{22} = 52/22 = 26/11$

$E(Y^2) = 1\dfrac{7}{22} + 9\dfrac{15}{22} = 142/22 = 71/11$

$\text{Cov}(XY) = 40/11 - (17/11)(26/11) = -2/121$

$\text{Var}(X) = 29/11 - (17/11)^2 = 30/121$

$\text{Var}(Y) = 71/11 - (26/11)^2 = 105/121$

$\rho(X,Y) = \dfrac{-2/121}{\sqrt{30/121}\sqrt{105/121}} = \dfrac{-2}{\sqrt{3150}} = \dfrac{-2}{15\sqrt{14}} = 0.036$

11.4.7 $E(X^2) = \displaystyle\int_0^1 x^2(4x^3)dx = 2/3.\ \ \text{Var}(X) = 2/3 - (4/5)^2 = 2/75$

$E(Y^2) = \displaystyle\int_0^1 y^2(4y - 4y^3)dx = 1/3.\ \ \text{Var}(Y) = 1/3 - (8/15)^2 = 11/225.$

$\rho = \dfrac{8/450}{\sqrt{2/75}\sqrt{11/225}} = 0.492$

11.4.9 $\rho(a + bX, c + dY) = \dfrac{\text{Cov}(a + bX, c + dY)}{\sqrt{\text{Var}(a + bX)\text{Var}(c + dY)}} = \dfrac{bd\,\text{Cov}(X,Y)}{\sqrt{b^2\text{Var}(X)d^2\text{Var}(Y)}}$, the equality in the

numerator stemming from Question 11.4.2. Since $b > 0, d > 0$, this last expression is

$\dfrac{bd\,\text{Cov}(X,Y)}{bd\sigma_X\sigma_Y} = \dfrac{\text{Cov}(X,Y)}{\sigma_X\sigma_Y} = \rho(X,Y).$

11.4.11 $\text{Cov}(X + Y, X - Y) = E[(X + Y)(X - Y)] - E(X + Y)E(X - Y)$

$= E[X^2 - Y^2] - (\mu_X + \mu_Y)(\mu_X - \mu_Y)$

$= E(X^2) - \mu_X - E(Y^2) + \mu_Y = \text{Var}(X) - \text{Var}(Y)$

11.4.13 Multiply the numerator and denominator of Equation 11.4.1 by n^2 to obtain

$$R = \dfrac{n\sum\limits_{i=1}^{n} X_i Y_i - \left(\sum\limits_{i=1}^{n} X_i\right)\left(\sum\limits_{i=1}^{n} Y_i\right)}{\sqrt{n\sum\limits_{i=1}^{n}(X_i - \overline{X})^2}\sqrt{n\sum\limits_{i=1}^{n}(Y_i - \overline{Y})^2}} = \dfrac{n\sum\limits_{i=1}^{n} X_i Y_i - \left(\sum\limits_{i=1}^{n} X_i\right)\left(\sum\limits_{i=1}^{n} Y_i\right)}{\sqrt{n\sum\limits_{i=1}^{n} X_i^2 - \left(\sum\limits_{i=1}^{n} X_i\right)^2}\sqrt{n\sum\limits_{i=1}^{n} Y_i^2 - \left(\sum\limits_{i=1}^{n} Y_i\right)^2}}$$

11.4.15 $r = \dfrac{n\sum\limits_{i=1}^{n}x_i y_i - \left(\sum\limits_{i=1}^{n}x_i\right)\left(\sum\limits_{i=1}^{n}y_i\right)}{\sqrt{n\sum\limits_{i=1}^{n}x_i^2 - \left(\sum\limits_{i=1}^{n}x_i\right)^2}\sqrt{n\sum\limits_{i=1}^{n}y_i^2 - \left(\sum\limits_{i=1}^{n}y_i\right)^2}}$

$= \dfrac{12(480{,}565) - (4936)(1175)}{\sqrt{12(3{,}071{,}116) - (4936)^2}\sqrt{12(123{,}349) - (1175)^2}} = -0.030.$ The data do not suggest that

altitude affects home run hitting.

11.4.17 $r = \dfrac{17(4{,}759{,}470) - (7{,}973)(8{,}517)}{\sqrt{17(4{,}611{,}291) - (7{,}973)^2}\sqrt{17(5{,}421{,}917) - (8{,}517)^2}} = 0.762.$ The amount of variation

attributed to the linear regression is $r^2 = (0.762)^2 = 0.581$, or 58.1%.

Section 11.5

11.5.1 Y is a normal random variable with $E(Y) = 6$ and $\text{Var}(Y) = 10$. Then $P(5 < Y < 6.5) =$

$P\left(\dfrac{5-6}{\sqrt{10}} < Z < \dfrac{6.5-6}{\sqrt{10}}\right) = P(-0.32 < Z < 0.16) = 0.5636 - 0.3745 = 0.1891.$ By Theorem

11.5.1, $Y\,|\,2$ is normal with $E(Y\,|\,2) = \mu_Y + \dfrac{\rho\sigma_Y}{\sigma_X}(2 - \mu_X) = 6 + \dfrac{\frac{1}{2}\sqrt{10}}{2}(2-3) = 5.209$

$\text{Var}(Y\,|\,2) = (1-\rho^2)\sigma_Y^2 = (1-0.25)10 = 7.5$, so the standard deviation of Y is $\sqrt{7.5} = 2.739$.

$P(5 < Y\,|\,2 < 6.5) = P\left(\dfrac{5-5.209}{2.739} < Z < \dfrac{6.5-5.209}{2.739}\right) = P(-0.08 < Z < 0.47) = 0.6808 - 0.4681$

$= 0.2127$

11.5.3 (a) $f_{X+Y}(t) = \dfrac{1}{2\pi\sqrt{1-\rho^2}}\int_{-\infty}^{\infty}\exp\left\{-\dfrac{1}{2}\left(\dfrac{1}{1-\rho^2}\right)\left[(t-y)^2 - 2\rho(t-y)y + y^2\right]\right\}dy$

The expression in the brackets can be expanded and rewritten as
$t^2 + 2(1+\rho)y^2 - 2t(1+\rho)y$
$= t^2 + 2(1+\rho)[y^2 - ty]$
$= t^2 + 2(1+\rho)\left[y^2 - ty + \dfrac{t^2}{4}\right] - \dfrac{1}{2}(1+\rho)t^2$
$= \dfrac{1-\rho}{2}t^2 + 2(1+\rho)(y-t/2)^2$. Placing this expression into the exponent gives

$f_{X+Y}(t) = \dfrac{1}{2\pi\sqrt{1-\rho^2}}e^{-\frac{1}{2}\left(\frac{1}{1-\rho^2}\right)\frac{1-\rho}{2}t^2}\int_{-\infty}^{\infty}e^{-\frac{1}{2}\left(\frac{1}{1-\rho^2}\right)2(1+\rho)(y-t/2)^2}dy$

$= f_{X+Y}(t) = \dfrac{1}{2\pi\sqrt{1-\rho^2}}e^{-\frac{1}{2}\left(\frac{t^2}{2(1+\rho)}\right)}\int_{-\infty}^{\infty}e^{-\frac{1}{2}\left(\frac{(y-t/2)^2}{(1+\rho)/2}\right)}dy.$

The integral is that of a normal pdf with mean $t/2$ and $\sigma^2 = (1+\rho)/2$.

Thus, the integral equals $\sqrt{2\pi(1+\rho)/2} = \sqrt{\pi(1+\rho)}$.

Putting this into the expression for f_{X+Y} gives

$$f_{X+Y}(t) = \frac{1}{\sqrt{2\pi}\sqrt{2(1+\rho)}}e^{-\frac{1}{2}\left(\frac{t^2}{2(1+\rho)}\right)}, \text{ which is the pdf of a normal variable with}$$

$\mu = 0$ and $\sigma^2 = 2(1+\rho)$.

(b) $E(X+Y) = c\mu_X + d\mu_Y$; $\text{Var}(X+Y) = c^2\sigma_X^2 + d^2\sigma_Y^2 + 2cd\sigma_X\sigma_Y\rho(X,Y)$

11.5.5 $E(X) = E(Y) = 0$; $\text{Var}(X) = 4$; $\text{Var}(Y) = 1$; $\rho(X, Y) = 1/2$; $k = 1/(2\pi\sqrt{3})$

11.5.7 $r = -0.453$. $T_{18} = \dfrac{\sqrt{n-2}\,r}{\sqrt{1-r^2}} = \dfrac{\sqrt{18}(-0.453)}{\sqrt{1-(-0.453)^2}} = -2.16$

Since $-t_{.005,18} = -2.8784 < T_{18} = -2.16 < 2.8784 = t_{.005,18}$, accept H_0.

11.5.9 From Question 11.4.15, $r = -0.030$. $T_{10} = \dfrac{\sqrt{10}(-0.030)}{\sqrt{1-(-0.030)^2}} = -0.09$.

Since $-t_{.025,10} = -2.2281 < T_{10} = -0.09 < 2.2281 = t_{.025,10}$, accept H_0.

11.5.11 From the Comment following Example 11.5.1, we can deduce that

$$P\left(-z_{\alpha/2} \le \frac{\frac{1}{2}\ln\frac{1+R}{1-R} - \frac{1}{2}\ln\frac{1+\rho}{1-\rho}}{\sqrt{\frac{1}{n-3}}} \le z_{\alpha/2}\right) = 1-\alpha$$

To find the confidence interval, we solve the inequality for ρ:

$$-z_{\alpha/2} \le \frac{\frac{1}{2}\ln\frac{1+R}{1-R} - \frac{1}{2}\ln\frac{1+\rho}{1-\rho}}{\sqrt{\frac{1}{n-3}}} \le z_{\alpha/2} \text{ implies}$$

$$-z_{\alpha/2}\sqrt{\frac{1}{n-3}} \le \frac{1}{2}\ln\frac{1+R}{1-R} - \frac{1}{2}\ln\frac{1+\rho}{1-\rho} \le z_{\alpha/2}\sqrt{\frac{1}{n-3}} \text{ or}$$

$$e^{-z_{\alpha/2}\sqrt{\frac{1}{n-3}}} \le \sqrt{\frac{1+R}{1-R}} \Big/ \sqrt{\frac{1+\rho}{1-\rho}} \le e^{z_{\alpha/2}\sqrt{\frac{1}{n-3}}}. \text{ Then}$$

$$\sqrt{\frac{1-R}{1+R}}\,e^{-z_{\alpha/2}\sqrt{\frac{1}{n-3}}} \le \sqrt{\frac{1-\rho}{1+\rho}} \le \sqrt{\frac{1-R}{1+R}}\,e^{z_{\alpha/2}\sqrt{\frac{1}{n-3}}}. \text{ Squaring the inequality gives}$$

$$\frac{1-R}{1+R}\,e^{-2z_{\alpha/2}\sqrt{\frac{1}{n-3}}} \le \frac{1-\rho}{1+\rho} \le \frac{1-R}{1+R}\,e^{2z_{\alpha/2}\sqrt{\frac{1}{n-3}}},$$

or $\dfrac{1-R}{1+R}e^{-2z_{\alpha/2}\sqrt{\frac{1}{n-3}}} \le -1 + \dfrac{2}{1+\rho} \le \dfrac{1-R}{1+R}e^{2z_{\alpha/2}\sqrt{\frac{1}{n-3}}}$. Solving this inequality for ρ yields the confidence interval:

$$1 + \dfrac{1-R}{1+R}e^{-2z_{\alpha/2}\sqrt{\frac{1}{n-3}}} \le \dfrac{2}{1+\rho} \le 1 + \dfrac{1-R}{1+R}e^{2z_{\alpha/2}\sqrt{\frac{1}{n-3}}}, \text{ which implies}$$

$$\dfrac{2}{1+\dfrac{1-R}{1+R}e^{2z_{\alpha/2}\sqrt{\frac{1}{n-3}}}} \le 1+\rho \le \dfrac{2}{1+\dfrac{1-R}{1+R}e^{-2z_{\alpha/2}\sqrt{\frac{1}{n-3}}}}, \text{ and finally}$$

$$-1 + \dfrac{2}{1+\dfrac{1-R}{1+R}e^{2z_{\alpha/2}\sqrt{\frac{1}{n-3}}}} < \rho < -1 + \dfrac{2}{1+\dfrac{1-R}{1+R}e^{-2z_{\alpha/2}\sqrt{\frac{1}{n-3}}}}$$

Chapter 12

Section 12.2

12.2.1 Here $n = 10$ and $k = 4$. To test H_0: $\mu_A = \mu_B = \mu_C = \mu_D$ at the $\alpha = 0.05$ level, reject the null hypothesis if $F \geq F_{.95, 4-1, 10-4} = 4.76$. At the $\alpha = 0.10$ level, H_0 is rejected if $F \geq F_{.90, 3, 6} = 3.29$. As the ANOVA table shows, the observed F falls between the two cutoffs, meaning that H_0 is rejected at the $\alpha = 0.10$ level, but not at the $\alpha = 0.05$ level.

Source	df	SS	MS	F
Model	3	61.33	20.44	3.94
Error	6	31.17	5.19	
Total	9	92.50		

12.2.3 For these $n = 30$ observations and $k = 3$ treatment groups, $C = T_{..}^2 / n = (422.9)^2 / 30 = 5961.48$, $\text{SSTOT} = \sum_{j=1}^{3} \sum_{i=1}^{10} y_{ij}^2 - C = 914.1$, and $\text{SSTR} = (121.4)^2/10 + (176.1)^2/10 + (125.4)^2/10 - 5961.48 = 186.0$. To test the null hypothesis that the three types of stocks have equal price-earnings ratios, reject H_0 if $F \geq F_{.99, 3-1, 30-3} = F_{.99, 2, 27}$. The latter is not a cutoff that appears in Table A.4 of the Appendix. However, its value can be bounded by cutoffs with similar degrees of freedom that are listed: $F_{.99, 2, 30} = 5.39 < F_{.99, 2, 27} < F_{.99, 2, 24} = 5.61$. According to the ANOVA table, the observed F ratio equals 3.45, which implies that H_0 should not be rejected.

Source	df	SS	MS	F
Sector	2	186.0	93.0	3.45
Error	27	728.2	27.0	
Total	29	914.1		

12.2.5 To test at the $\alpha = 0.01$ level of significance the null hypothesis that the four tribes were contemporaries of one another, H_0 should be rejected if $F \geq F_{.99, 4-1, 12-4} = 7.59$ (or if $P < 0.01$). According to the ANOVA table, F is less than 7.59 (and P is greater than 0.01), so H_0 should not be rejected.

Source	df	SS	MS	F	P
Tribe	3	504167	168056	3.70	0.062
Error	8	363333	45417		
Total	11	867500			

12.2.7

Source	df	SS	MS	F
Treatment	4	271.36	67.84	6.40
Error	10	106.00	10.60	
Total	14	377.36		

12.2.9 $\text{SSTOT} = \sum\limits_{j=1}^{k}\sum\limits_{i=1}^{n_j}(Y_{ij} - \bar{Y}_{..})^2 = \sum\limits_{j=1}^{k}\sum\limits_{i=1}^{n_j}(Y_{ij}^2 - 2Y_{ij}\bar{Y}_{..} + \bar{Y}_{..}^2) =$

$$\sum_{j=1}^{k}\sum_{i=1}^{n_j}Y_{ij}^2 - 2\bar{Y}_{..}\sum_{j=1}^{k}\sum_{i=1}^{n_j}Y_{ij} + n\bar{Y}_{..}^2 = \sum_{j=1}^{k}\sum_{i=1}^{n_j}Y_{ij}^2 - 2n\bar{Y}_{..}^2 + n\bar{Y}_{..}^2 =$$

$$\sum_{j=1}^{k}\sum_{i=1}^{n_j}Y_{ij}^2 - n\bar{Y}_{..}^2 = \sum_{j=1}^{k}\sum_{i=1}^{n_j}Y_{ij}^2 - C, \text{ where } C = T_{..}^2/n. \text{ Also,}$$

$$\text{SSTR} = \sum_{j=1}^{k}\sum_{i=1}^{n_j}(\bar{Y}_{.j}^2 - \bar{Y}_{..})^2 = \sum_{j=1}^{k}n_j(\bar{Y}_{.j}^2 - 2\bar{Y}_{.j}\bar{Y}_{..} + \bar{Y}_{..}^2) =$$

$$\sum_{j=1}^{k}T_{.j}^2/n_j - 2\bar{Y}_{..}\sum_{j=1}^{k}n_j\bar{Y}_{.j} + n\bar{Y}_{..}^2 = \sum_{j=1}^{n}T_{.j}^2/n_j - 2n\bar{Y}_{..}^2 + n\bar{Y}_{..}^2 =$$

$$\sum_{j=1}^{k}T_{.j}^2/n_j - C.$$

12.2.11 Analyzed with a two-sample t test, the data in Question 9.2.4 require that H_0: $\mu_X = \mu_Y$ be rejected (in favor of a two-sided H_1) at the $\alpha = 0.05$ level if $|t| \geq t_{.025, 6+9-2} = 2.1604$. Evaluating the test statistic gives $t = (70.83 - 79.33)/11.31\sqrt{1/6 + 1/9} = -1.43$, which implies that H_0 should not be rejected. The ANOVA table for the same data shows that $F = 2.04$. But $(-1.43)^2 = 2.04$. Moreover, H_0 would be rejected with the analysis of variance if $F \geq F_{.95, 1, 13} = 4.667$. But $(2.1604)^2 = 4.667$.

Source	df	SS	MS	F
Sex	1	260	260	2.04
Error	13	1661	128	
Total	14	1921		

12.2.13

Source	df	SS	MS	F	P
Law	1	16.333	16.333	1.58	0.2150
Error	46	475.283	10.332		
Total	47	491.617			

The F critical value is 4.05.

For the pooled two-sample t test, the observed t ratio is -1.257, and the critical value is 2.1029.

Note that $(-1.257)^2 = 1.58$ (rounded to two decimal places) which is the observed F ratio. Also, $2.0129^2 = 4.05$ (rounded to two decimal places), which is the F critical value.

Section 12.3

12.3.1 For the data in Case Study 12.2.1, $k = 4$, $r = 6$, MSE = 79.74, and $D = Q_{.05,4,20}/\sqrt{6} =$
$3.96/\sqrt{6} = 1.617$. Let μ_1, μ_2, μ_3, and μ_4 denote the true average heart rates for Non-smokers, Light smokers, Moderate smokers, and Heavy smokers, respectively. Substituting into Theorem 12.3.1 gives the six different Tukey intervals summarized in the table below.

Pairwise Difference	Tukey interval	Conclusion
$\mu_1 - \mu_2$	$(-15.27, 13.60)$	NS
$\mu_1 - \mu_3$	$(-23.77, 5.10)$	NS
$\mu_1 - \mu_4$	$(-33.77, -4.90)$	Reject
$\mu_2 - \mu_3$	$(-22.94, 5.94)$	NS
$\mu_2 - \mu_4$	$(-32.94, -4.06)$	Reject
$\mu_3 - \mu_4$	$(-24.44, 4.44)$	NS

12.3.3 Let μ_C, μ_A, and μ_M denote the true average numbers of contaminant particles in IV fluids produced by Cutter, Abbott, and McGaw, respectively. According to the analysis of variance, H_0: $\mu_C = \mu_A = \mu_M$ is rejected at the $\alpha = 0.05$ level (since the P value is less than

Source	df	SS	MS	F	P
Company	2	113646	56823	5.81	0.014
Error	15	146754	9784		
Total	17	260400			

0.05). The three 95% Tukey confidence intervals (based on $k = 3$, $r = 6$, and $D = Q_{.05,3,15}/\sqrt{6} =$
$3.67/\sqrt{6} = 1.498$) show that Abbott and McGaw have the only pairwise difference
$(204.50 - 396.67 = -192.17)$ that is statistically significant.

Pairwise Difference	Tukey interval	Conclusion
$\mu_C - \mu_A$	$(-78.9, 217.5)$	NS
$\mu_C - \mu_M$	$(-271.0, 25.4)$	NS
$\mu_A - \mu_M$	$(-340.0, -44.0)$	Reject

12.3.5 Since $k = 3$ and $r = 3$, $D = Q_{.05,3,6}/\sqrt{3} = 4.34/\sqrt{3} = 2.506$. MSE = 41.666, so the radius of the intervals is $D\sqrt{\text{MSE}} = 2.506\sqrt{41.667} = 16.17$.

Pairwise Difference	Tukey interval	Conclusion
$\mu_1 - \mu_2$	$(-29.5, 2.8)$	NS
$\mu_1 - \mu_3$	$(-56.2, -23.8)$	Reject
$\mu_2 - \mu_3$	$(-42.8, -10.5)$	REject

12.3.7 Longer. As k gets larger, the number of possible pariwise comparisons increases. To maintain the same overall probability of committing at least one Type I error, the individual intervals would need to be widened.

Section 12.4

12.4.1

Source	df	SS	MS	F
Tube	2	510.7	255.4	11.56
Error	42	927.7	22.1	
Total	44	1438.4		

Subhypothesis	Contrast	SS	F
$H_0:\ \mu_A = \mu_C$	$C_1 = \mu_A - \mu_C$	264.0	11.95
$H_0:\ \mu_B = \dfrac{\mu_A + \mu_C}{2}$	$C_2 = \dfrac{1}{2}\mu_A - \mu_B + \dfrac{1}{2}\mu_C$	246.7	11.16

$H_0:\ \mu_A = \mu_B = \mu_C$ is strongly rejected ($F_{.99,2,42} \doteq F_{.99,2,40} = 5.18$). Theorem 12.4.1 holds true for orthogonal contrasts C_1 and C_2—$SS_{C_1} + SS_{C_2} = 264.0 + 246.7 = 510.7 = SSTR$. Also, both subhypotheses would be rejected—their F ratios exceed $F_{.99,1,42}$.

12.4.3 Let μ_1, μ_2, μ_3, and μ_4 denote the true average heart rates for Non-smokers, Light smokers, Moderate smokers, and Heavy smokers, respectively. To test $H_0:\ (\mu_2 + \mu_3)/2 = \mu_4$, let $C = \dfrac{1}{2}\mu_2 + \dfrac{1}{2}\mu_3 - \mu_4$, so $\hat{C} = \dfrac{1}{2}(63.2) + \dfrac{1}{2}(71.7) - 1(81.7) = -14.25$. Also, $SS_C =$

$$(-14.25)^2 / \left[\frac{(1/2)^2}{6} + \frac{(1/2)^2}{6} + \frac{(-1)^2}{6} \right] = 812.25.$$ From the ANOVA table on p. 642, $SSE = 1594.833$ and $n - k = 20$. Therefore, H_0 should be rejected if $F \geq F_{.95,1,20} = 4.35$. Here $F = \dfrac{812.25/1}{1594.833/20} = 10.19$, so $H_0:\ (\mu_2 + \mu_3)/2 = \mu_4$ is rejected (at the $\alpha = 0.05$ level).

12.4.5

	μ_A	μ_B	μ_C	μ_D	$\sum_{j=1}^{4} c_j$
C_1	1	-1	0	0	0
C_2	0	0	1	-1	0
C_3	$\dfrac{11}{12}$	$\dfrac{11}{12}$	-1	$\dfrac{-5}{6}$	0

C_1 and C_3 are orthogonal because $\dfrac{1(11/12)}{6} + \dfrac{(-1)(11/12)}{6} = 0$; also, C_2 and C_3 are orthogonal because $\dfrac{1(-1)}{6} + \dfrac{(-1)(-5/6)}{5} = 0$. $\hat{C}_3 = -2.293$ and $SS_{C_3} = 8.97$. But $SS_{C_1} + SS_{C_2} + SS_{C_3} = 4.68 + 1.12 + 8.97 = 14.77 = SSTR$.

Section 12.5

12.5.1 Replace each observation by its square root. At the $\alpha = 0.05$ level, H_0: $\mu_A = \mu_B$ is rejected. (For $\alpha = 0.01$, though, we would fail to reject H_0).

Source	df	SS	MS	F	P
Developer	1	1.836	1.836	6.23	0.032
Error	10	2.947	0.295		
Total	11	4.783			

12.5.3 Since Y_{ij} is a binomial random variable based on $n = 20$ trials, each data point should be replaced by the arcsin of $(y_{ij}/20)^{\frac{1}{2}}$. Based on those transformed observations, H_0: $\mu_A = \mu_B = \mu_C$ is strongly rejected ($P < 0.001$).

Source	df	SS	MS	F	P
Launcher	2	0.30592	0.15296	22.34	0.000
Error	9	0.06163	0.00685		
Total	11	0.36755			

Appendix 12.A.3

12.A.3.1 The F test will have greater power against H_1^{**} because the latter yields a larger noncentrality parameter than does H_1^*.

12.A.3.3 $M_V(t) = (1 - 2t)^{-r/2} e^{\gamma t (1-2t)^{-1}}$, so $M_V^{(1)}(t) = (1-2t)^{-r/2} \cdot e^{\gamma t (1-2t)^{-1}} [\gamma t(-1)(1-2t)^{-2}(-2) + (1-2t)^{-1}\gamma] + e^{\gamma t(1-2t)^{-1}} \left(-\dfrac{r}{2}\right)(1-2t)^{-(r/2)-1}(-2)$. Therefore, $E(V) = M_V^{(1)}(0) = \gamma + r$.

12.A.3.5 $M_V(t) = \displaystyle\prod_{i=1}^{n}(1-2t)^{-r_i/2} e^{\gamma_i t/(1-2t)} = (1-2t)^{-\sum_{i=1}^{n} r_i/2} \cdot e^{\left(\sum_{i=1}^{n}\gamma_i\right) t/(1-2t)}$, which implies that V has a noncentral χ^2 distribution with $\displaystyle\sum_{i=1}^{n} r_i$ df and with noncentrality parameter $\displaystyle\sum_{i=1}^{n}\gamma_i$.

Chapter 13

Section 13.2

13.2.1

Source	df	SS	MS	F	P
States	1	61.63	61.63	7.20	0.0178
Students	14	400.80	28.63	3.34	0.0155
Error	14	119.87	8.56		
Total	29	582.30			

The critical value $F_{.95,1,14}$ is approximately 4.6. Since the F statistic $= 7.20 > 4.6$, reject H_0.

13.2.3

Source	df	SS	MS	F	P
Additive	1	0.03	0.03	4.19	0.0865
Batch	6	0.02	0.00	0.41	0.8483
Error	6	0.05	0.01		
Total	13	0.10			

Since the F statistic $= 4.19 < F_{.95,1,6} = 5.99$, accept H_0.

13.2.5

Source	df	SS	MS	F	P
Quarter	3	0.60	0.20	0.60	0.6272
Year	4	19.87	4.97	14.85	0.0001
Error	12	4.01	0.33		
Total	19	24.48			

Since the F statistic for treatments $= 0.60 < F_{.95,3,12} = 3.49$, accept H_0 that yields are not affected by the quarter.
Since the F statistic for blocks $= 14.85 > F_{.95,4,12} = 3.26$, reject H_0; yields do depend on the year.

13.2.7 From Question 13.2.2 we obtain the value MSE $= 0.98$. The radius of the interval is
$$D\sqrt{\text{MSE}} = (Q_{.05,3,6}/\sqrt{b})\sqrt{0.98} = (4.34/\sqrt{4})\sqrt{0.98} = 2.148.$$

The Tukey intervals are

Pairwise Difference	$\bar{y}_{.s} - \bar{y}_{.t}$	Tukey Interval	Conclusion
$\mu_1 - \mu_2$	2.925	(0.78, 5.07)	Reject
$\mu_1 - \mu_3$	1.475	(−0.67, 3.62)	NS
$\mu_2 - \mu_3$	−1.450	(−3.60, 0.70)	NS

13.2.9 (a)

Source	df	SS	MS	F	P
Sleep phases	2	16.99	8.49	4.13	0.0493
Shrew	5	195.44	39.09	19.00	0.0001
Error	10	20.57	2.06		
Total	17	233.00			

Since the F statistic $= 4.13 > F_{.95, 2, 10} = 4.10$, reject H_0.

(b)   The contrast associated with the subhypothesis is

$$C_1 = -\frac{1}{2}\mu_1 - \frac{1}{2}\mu_2 + \mu_3, \text{ and } \hat{C}_1 = -\frac{1}{2}(21.1) - \frac{1}{2}(19.1) + 18.983 = -1.117$$

$$SS_{C_1} = \frac{(-1.117)^2}{\left(-\frac{1}{2}\right)^2/6 + \left(-\frac{1}{2}\right)^2/6 + (1)^2/6} = 4.99.$$

$$F = \frac{SS_{C_1}}{MSE} = \frac{4.99}{2.06} = 2.42. \text{ Since the observed } F \text{ ratio} = 2.42 < F_{.95, 1, 10} = 4.96, \text{ accept}$$
the subhypothesis.

Let a second orthogonal contrast be $C_2 = \mu_1 - \mu_2$.

$$\hat{C}_2 = 21.1 - 19.1 = 2.0. \quad SS_{C_2} = \frac{2.0^2}{(1)^2/6 + (-1)^2/6} = 12.0$$

Then SSTR $= 16.99 = 4.99 + 12.00 = SS_{C_1} + SS_{C_2}$

13.2.11 Equation 13.2.2:

$$SSTR = \sum_{i=1}^{b}\sum_{j=1}^{k}(\bar{Y}_{\cdot j} - \bar{Y})^2 = b\sum_{j=1}^{k}(\bar{Y}_{\cdot j} - \bar{Y}..)^2$$

$$= b\sum_{j=1}^{k}(\bar{Y}_{\cdot j}^2 - 2\bar{Y}_{\cdot j}\bar{Y}.. + \bar{Y}..^2) = b\sum_{j=1}^{k}\bar{Y}_{\cdot j}^2 - 2b\bar{Y}..\sum_{j=1}^{k}\bar{Y}_{\cdot j} + bk\bar{Y}..^2$$

$$= b\sum_{j=1}^{k}\frac{T_{\cdot j}^2}{b^2} - \frac{2T..^2}{bk} + \frac{T..^2}{bk} = \sum_{j=1}^{k}\frac{T_{\cdot j}^2}{b} - \frac{T..^2}{bk} = \sum_{j=1}^{k}\frac{T_{\cdot j}^2}{b} - c$$

Equation 13.2.3:

$$SSB = \sum_{i=1}^{b}\sum_{j=1}^{k}(\bar{Y}_{i\cdot} - \bar{Y})^2 = k\sum_{i=1}^{b}(\bar{Y}_{i\cdot} - \bar{Y})^2$$

$$= k\sum_{i=1}^{b}(\bar{Y}_{i\cdot}^2 - 2\bar{Y}_{i\cdot}\bar{Y}.. + \bar{Y}..^2) = k\sum_{i=1}^{b}\bar{Y}_{i\cdot}^2 - 2k\bar{Y}..\sum_{i=1}^{b}\bar{Y}_{i\cdot} + bk\bar{Y}..^2$$

$$= k\sum_{i=1}^{b}\frac{T_{i\cdot}^2}{k^2} - \frac{2T..^2}{bk} + \frac{T..^2}{bk} = \sum_{i=1}^{b}\frac{T_{i\cdot}^2}{k} - \frac{T..^2}{bk} = \sum_{i=1}^{b}\frac{T_{i\cdot}^2}{k} - c$$

Chapter 13   **111**

Equation 13.2.4:

$$\text{SSTOT} = \sum_{i=1}^{b}\sum_{j=1}^{k}(Y_{ij} - \overline{Y}_{..})^2 = \sum_{i=1}^{b}\sum_{j=1}^{k}(Y_{ij}^2 - 2Y_{ij}\overline{Y}_{..} + \overline{Y}_{..}^2)$$

$$\sum_{i=1}^{b}\sum_{j=1}^{k}Y_{ij}^2 - 2\overline{Y}_{..}\sum_{i=1}^{b}\sum_{j=1}^{k}Y_{ij} + bk\overline{Y}_{..}^2$$

$$= \sum_{i=1}^{b}\sum_{j=1}^{k}Y_{ij}^2 - \frac{2T_{..}^2}{bk} + \frac{T_{..}^2}{bk} = \sum_{i=1}^{b}\sum_{j=1}^{k}Y_{ij}^2 - c$$

13.2.13 (a) False. $\displaystyle\sum_{i=1}^{b}\overline{Y}_{i.} = \frac{1}{k}\sum_{i=1}^{b}\sum_{j=1}^{k}Y_{ij}$. $\displaystyle\sum_{j=1}^{k}\overline{Y}_{.j} = \frac{1}{b}\sum_{i=1}^{b}\sum_{j=1}^{k}Y_{ij}$. The two expressions are equal only when $b = k$.

(b) False. If neither treatment levels nor blocks are significant, it is possible to have F variables $\dfrac{\text{SSTR}/(k-1)}{\text{SSE}/(b-1)(k-1)}$ and $\dfrac{\text{SSB}/(b-1)}{\text{SSE}/(b-1)(k-1)}$ both < 1.
In that case both SSTR and SSB are less than SSE.

Section 13.3

13.3.1 Test H_0: $\mu_D = 0$ vs H_1: $\mu_D > 0$.

$$s_D^2 = \frac{b\sum_{i=1}^{b}d_i^2 - \left(\sum_{i=1}^{b}d_i\right)^2}{b(b-1)} = \frac{12(11.7229) - (8.55)^2}{12(11)} = 0.512$$

$$t = \frac{\overline{d}}{s_D/\sqrt{b}} = \frac{0.7125}{\sqrt{0.512}/\sqrt{12}} = 3.45$$

Since $3.45 > 1.7959 = t_{.05,11}$, reject H_0.

13.3.3 Test H_0: $\mu_D = 0$ vs. H_1: $\mu_D \neq 0$.

$$s_D^2 = \frac{b\sum_{i=1}^{b}d_i^2 - \left(\sum_{i=1}^{b}d_i\right)^2}{b(b-1)} = \frac{12(2.97) - (1.3)^2}{12(11)} = 0.257$$

$$t = \frac{\overline{d}}{s_D/\sqrt{b}} = \frac{1.108}{\sqrt{0.257}/\sqrt{12}} = 0.74.$$

$\alpha = 0.05$: Since $-t_{.025,11} = -2.2010 < 0.74 < 2.2010 = t_{.025,11}$, accept H_0.
$\alpha = 0.01$: Since $-t_{.005,11} = -3.1058 < 0.74 < 3.1058 = t_{.005,11}$ accept H_0.

13.3.5 Test H_0: $\mu_D = 0$ vs. H_1: $\mu_D \neq 0$.

$$s_D^2 = \frac{7(0.1653) - (-0.69)^2}{7(6)} = 0.01621$$

$$t = \frac{-0.09857}{\sqrt{0.01621}/\sqrt{7}} = -2.048.$$

Since $-t_{.025,6} = -2.4469 < -2.048 < 2.4469 = t_{.025,6}$ accept H_0.

The square of the observed Student t statistic $= (-2.048)^2 = 4.194 =$ the observed F statistic.

Also, $(t_{.025,6})^2 = (2.4469)^2 = 5.987 = F_{.95,1,6}$

Conclusion: the two-sided test for paired data is equivalent to the randomized block design test for 2 treatments.

13.3.7 The 95% confidence interval is $\left(\bar{d} - t_{.025,11} \frac{s_D}{\sqrt{b}}, \bar{d} + t_{.025,11} \frac{s_D}{\sqrt{b}} \right)$

$$= \left(0.108 - 2.2010 \frac{\sqrt{0.257}}{\sqrt{12}}, 0.108 + 2.2010 \frac{\sqrt{0.257}}{\sqrt{12}} \right) = (-0.21, 0.43)$$

Chapter 14

Section 14.2

14.2.1 Here $x = 8$ of the $n = 10$ groups were larger than the hypothesized median of 9. The P-value is

$$P(X \geq 8) + P(X \leq 2) = 0.000977 + 0.009766 + 0.043945 + 0.043945 + 0.009766 + 0.000977$$
$$= 2(0.054688) = 0.109376$$

14.2.3 The median of $f_Y(y)$ is 0.693. There are $x = 22$ values that exceed the hypothesized median of 0.693. The test statistic is $z = \dfrac{22 - 50/2}{\sqrt{50/4}} = -0.85$.

Since $-z_{0.025} = -1.96 < -0.85 < z_{0.025} = 1.96$, do not reject H_0.

14.2.5 $P(Y_+ = y_+) = \dbinom{7}{y_+}\dfrac{1}{2^7}$. These values are given in the table.

y_+	$P(Y_+ = y_+)$
0	1/128
1	7/128
2	21/128
3	35/128
4	35/128
5	21/128
6	7/128
7	1/128

Possible levels for a one-sided test: 1/128, 8/128, 29/128, etc.

14.2.7

y_i	$y_i - 0.80$	sign	y_i	$y_i - 0.80$	sign
0.61	−0.19	−	0.78	−0.02	−
0.70	−0.10	−	0.84	0.04	+
0.63	−0.17	−	0.83	0.03	+
0.76	−0.04	−	0.82	0.02	+
0.67	−0.13	−	0.74	−0.06	−
0.72	−0.08	−	0.85	0.05	+
0.64	−0.16	−	0.73	−0.07	−
0.82	0.02	+	0.85	0.05	+
0.88	0.08	+	0.87	0.07	+
0.82	0.02	+			

$$\sum_{k=0}^{6}\binom{19}{k}\left(\frac{1}{2}\right)^{19} = 0.0835, \text{ while } \sum_{k=0}^{7}\binom{19}{k}\left(\frac{1}{2}\right)^{19} = 0.1796. \text{ Thus, the closest test to one with } \alpha$$

$= 0.10$ is to reject H_0 if $y_+ \leq 6$. This test has $\alpha = 0.0835$. Since $y_+ = 9$, accept H_0.
Since the observed t statistic $= -1.71 < -1.330 = -t_{.10,18}$, the t test rejects H_0.

14.2.9 $z = \dfrac{y_+ - \frac{1}{2}n}{\sqrt{n/4}} = \dfrac{19 - \frac{1}{2}(28)}{\sqrt{28/4}} = 1.89.$ Accept H_0, since $-z_{.025} = -1.96 < 1.89 < 1.96 = z_{.025}$.

14.2.11 From Table 13.3.1, the number of pairs where $x_i > y_i$ is 7. The P-value for this test is $P(U \geq 7) + P(U \leq 3) = 2(0.17186) = 0.343752$. Since the P value exceeds $\alpha = 0.05$, do not reject the null hypothesis, which is the conclusion of Case Study 13.3.1.

Section 14.3

14.3.1

x_i	y_i	$y_i - x_i$	$|y_i - x_i|$	r_i	z_i	$r_i z_i$
1458	1424	−34	34	1	0	0
1353	1501	148	148	5	1	5
2209	1495	−714	714	8	0	0
1804	1739	−65	65	2	0	0
1912	2031	119	119	4	1	4
1366	934	−432	432	7	0	0
1598	1401	−197	197	6	0	0
1406	1339	−67	67	3	0	0

The sum of the $r_i z_i$ column is 9. From Table A.6, the critical values of 7 and 29 give $\alpha = 0.148$. Since $7 < w = 9 < 29$, accept H_0.

14.3.3

x_i	y_i	$y_i - x_i$	$|y_i - x_i|$	r_i	z_i	$r_i z_i$
16.5	16.9	0.4	0.4	12.5	1	12.5
17.6	17.2	−0.4	0.4	12.5	0	0
16.9	17.0	0.1	0.1	2	1	2
15.8	16.1	0.3	0.3	8.5	1	8.5
18.4	18.2	−0.2	0.2	4.5	0	0
17.5	17.7	0.2	0.2	4.5	1	4.5
17.6	17.9	0.3	0.3	8.5	1	8.5
16.1	16.0	−0.1	0.1	2	0	0
16.8	17.3	0.5	0.5	14	1	14
15.8	16.1	0.3	0.3	8.5	1	8.5
16.8	16.5	−0.3	0.3	8.5	0	0
17.3	17.6	0.3	0.3	8.5	1	8.5
18.1	18.4	0.3	0.3	8.5	1	8.5
17.9	17.2	−0.7	0.7	15	0	0
16.4	16.5	0.1	0.1	2	1	2

$w = $ sum of the $r_1 z_i$ column is 77.5. The mean of W is $n(n+1)/4 = 60$. The variance of $W = n(n+1)(2n+1)/24 = 310$. The observed Z statistic $w' = \dfrac{77.5 - 60}{\sqrt{310}} = 0.99$.

Since $-1.96 < w' = 0.99 < 1.96 = z_{.025}$, accept H_0.

y_i	$y_i - 0.80$	$\lvert y_i - 0.80 \rvert$	r_i	z_i	$r_i z_i$
0.61	−0.19	0.19	19	0	0
0.70	−0.10	0.10	15	0	0
0.63	−0.17	0.17	18	0	0
0.76	−0.04	0.04	6.5	0	0
0.67	−0.13	0.13	16	0	0
0.72	−0.08	0.08	13.5	0	0
0.64	−0.16	0.16	17	0	0
0.82	0.02	0.02	2.5	1	2.5
0.88	0.08	0.08	13.5	1	13.5
0.82	0.02	0.02	2.5	1	2.5
0.78	−0.02	0.02	2.5	0	0
0.84	0.04	0.04	6.5	1	6.5
0.83	0.03	0.03	5	1	5
0.82	0.02	0.02	2.5	1	2.5
0.74	−0.06	0.06	10	0	0
0.85	0.05	0.05	8.5	1	8.5
0.73	−0.07	0.07	11.5	0	0
0.85	0.05	0.05	8.5	1	8.5
0.87	0.07	0.07	11.5	1	11.5

14.3.5

w = sum of the $r_i z_i$ column = 61. The mean of W is $n(n + 1)/4 = 95$. The variance of $W = n(n + 1)(2n + 1)/24 = 617.5$. The observed Z statistic $w' = \dfrac{61 - 95}{\sqrt{617.5}} = -1.37$. Since $w' = -1.37 < -1.28 = -z_{.10}$, reject H_0. The sign test accepted H_0.

14.3.7 The signed rank test should have more power since it uses a greater amount of the information in the data.

14.3.9 A reasonable assumption is that alcohol abuse shortens life span. In that case, reject H_0 if the test statistic is less than $-z_{.05} = -1.64$. From the table below, we see that $w' = 72.5$. The test statistic is $z = \dfrac{72.5 - 99}{\sqrt{198}} = -1.88$, so reject H_0.

Age at death	r_i	z_i	$r_i z_i$
48	2	1	2
66	7	1	7
71	12	1	12
65	5.5	1	5.5
56	3	1	3
67	9	1	9
67	9	1	9
70	11	1	11
77	14	1	14
65	5.5	0	0
87	18	0	0
32	1	0	0
77	14	0	0
89	20	0	0
86	17	0	0
77	14	0	0
84	16	0	0
64	4	0	0
88	19	0	0
90	21	0	0
67	9	0	0
			$w' = 72.5$

Section 14.4

14.4.1

Group I	Rank	Group II	Rank	Group III	Rank
3	2.5	10	9	20	15
2	1	4	4	9	7
6	6	11	11	18	13
10	9	14	12	19	14
10	9	3	2.5		
5	5				

The sum of the second column in the table is $r_{.1} = 32.5$; the fourth column, $r_{.2} = 38.5$; and the sixth column $r_{.3} = 49$. The test statistic is

$$b = \frac{12}{n(n+1)}\left(\frac{r_{.1}^2}{n_1} + \frac{r_{.2}^2}{n_2} + \frac{r_{.3}^2}{n_3}\right) - 3(n+1) = \frac{12}{15(16)}\left(\frac{32.5^2}{6} + \frac{38.5^2}{5} + \frac{49^2}{4}\right) - 3(16) = 5.64.$$

Since $5.64 < 5.991 = \chi_{.95,2}$, accept H_0.

14.4.3	Amer. Male	Rank	Eur. Male	Rank
	14.6	9	3.6	2
	28.8	16	8.2	5
	19.1	12	7.8	4
	23.1	14	27.5	15
	50.3	18	7.0	3
	35.7	17	19.7	13
			17.0	10
			3.5	1
			13.3	7
			12.4	6
			19.0	11
			14.1	8

Summing the second column in the table gives $r_{.1} = 86$, and the sum of the fourth column is $r_{.2} = 85$. The test statistic is $b = \dfrac{12}{n(n+1)}\left(\dfrac{r_{.1}^2}{n_1} + \dfrac{r_{.2}^2}{n_2}\right) - 3(n+1)$

$$= \dfrac{12}{18(19)}\left(\dfrac{86^2}{8} + \dfrac{85^2}{12}\right) - 3(19) = 7.38.$$

Since $b = 7.38 > 3.841 = \chi^2_{.95,1}$, reject H_0.

14.4.5	Non-Smokers	Rank	Light Smokers	Rank	Moderate Smokers	Rank	Heavy Smokers	Rank
	69	13	55	2	66	10.5	91	23
	52	1	60	7	81	20.5	72	16
	71	15	78	18	70	14	81	20.5
	58	4.5	58	4.5	77	17	67	12
	59	6	62	8	57	3	95	24
	65	9	66	10.5	79	19	84	22

$$b = \dfrac{12}{n(n+1)}\left(\dfrac{r_{.1}^2}{n_1} + \dfrac{r_{.2}^2}{n_2} + \dfrac{r_{.3}^2}{n_3} + \dfrac{r_{.4}^2}{n_4}\right) - 3(n+1)$$

$$= \dfrac{12}{24(25)}\left(\dfrac{48.5^2}{6} + \dfrac{50^2}{6} + \dfrac{84^2}{6} + \dfrac{117.5^2}{6}\right) - 3(25) = 10.715$$

Since $b = 10.715 > 7.815 = \chi^2_{.95,3}$, reject H_0.

14.4.7

Powdered	Rank	Moderate	Rank	Coarse	Rank
146	8.5	150	14.5	141	4
152	16	144	6	138	2
149	12.5	148	10.5	142	5
161	21	155	19	146	8.5
158	20	154	17.5	139	3
149	12.5	150	14.5	145	7
154	17.5	148	10.5	137	1

$$b = \frac{12}{n(n+1)}\left(\frac{r_1^2}{n_1} + \frac{r_2^2}{n_2} + \frac{r_3^2}{n_3}\right) - 3(n+1)$$

$$= \frac{12}{21(22)}\left(\frac{108^2}{7} + \frac{92.5^2}{7} + \frac{30.5^2}{7}\right) - 3(22) = 12.48$$

Since $b = 12.48 > 5.991 = \chi_{.95,2}^2$, reject H_0.

Section 14.5

14.5.1

36 lb.	Rank	54 lb.	Rank	72 lb.	Rank	108 lb.	Rank	144 lb.	Rank
7.62	3	8.14	5	7.76	4	7.17	1	7.46	2
8.00	4	8.15	5	7.73	3	7.57	1	7.68	2
7.93	5	7.87	4	7.74	2	7.80	3	7.21	1

$$g = \frac{12}{bk(k+1)}\sum_{j=1}^{5} r_j^2 - 3b(k+1)$$

$$= \frac{12}{3(5)(6)}(12^2 + 14^2 + 9^2 + 5^2 + 5^2) - 3(3)(6)$$

Since $g = 8.8 < 9.488 = \chi_{.95,4}^2$, accept H_0.

14.5.3

PcrCh1	Rank	Davies	Rank	AOAC	Rank
0.598	1	0.628	2	0.632	3
0.614	1	0.628	2	0.630	3
0.600	1.5	0.600	1.5	0.622	3
0.580	1	0.612	3	0.584	2
0.596	1	0.600	2	0.650	3
0.592	1	0.628	3	0.606	2
0.616	1	0.628	2	0.644	3
0.614	1	0.644	2.5	0.644	2.5
0.604	1	0.644	3	0.624	2
0.608	1	0.612	2	0.619	3
0.602	1	0.628	2	0.632	3
0.614	1	0.644	3	0.616	2

$$g = \frac{12}{bk(k+1)} \sum_{j=1}^{3} r_{\cdot j}^2 - 3b(k+1)$$

$$= \frac{12}{12(3)(4)}(12.5^2 + 28^2 + 31.5^2) - 3(12)(4) = 17.0$$

Since $g = 17.0 > 5.991 = \chi_{.95,2}^2$ reject H_0.

14.5.5

Contact Desensitization	Rank	Demonstration Participation	Rank	Live Modeling	Rank
8	3	2	2	−2	1
11	3	1	2	0	1
9	2	12	3	6	1
16	3	11	2	2	1
24	3	19	2	11	1

$$g = \frac{12}{5(3)(4)}\left(14^2 + 11^2 + 5^2\right) - 3(5)(4) = 8.4$$

Since $g = 8.4 < 9.210 = \chi_{.99,2}$, accept H_0. On the other hand, using analysis of variance, the null hypothesis would be rejected at this level.

Section 14.5

14.6.1 (a)

% Change January	$\text{sgn}(y_i - y_{i-1})$	% Change January	$\text{sgn}(y_i - y_{i-1})$
2.0	+	−2.9	+
2.3	−	0.7	−
0.6	−	0.0	+
−0.9	+	1.4	+
0.5	−	1.5	−
−1.8	−	−1.5	+
−2.1	+	2.2	+
−0.9	+	4.9	−
2.5	−	−2.3	−
0.3	−	−4.6	+
−0.7	+	2.8	−
1.2	−	0.9	−
−3.4	+	−2.0	−
2.6	−	−2.4	+
1.3	−	3.2	−
0.7	+	2.4	−
0.8	+	−1.9	+
3.1	−	−1.6	
0.2	−		

For these data, the number of runs $w = 23$. The test statistic is $Z = \dfrac{W - E(W)}{\sqrt{\text{Var}(W)}}$, where $E(W) = (2n - 1)/3$ and $\text{Var}(W) = (16n - 29)/90$. For these data, $E(W) = 24.33$ and $\text{Var}(W) = 6.26$. Then $z = \dfrac{23 - 24.33}{\sqrt{6.26}} = -0.53$. Since $-z_{.025} = -1.96 < -0.53 < 1.96 = z_{.025}$, accept H_0 and assume the sequence is random.

(b) The number of runs is $w = 21$ and $z = \dfrac{21 - 24.33}{\sqrt{6.26}} = -1.33$.

Since $-z_{.025} = -1.96 < -1.33 < 1.96 = z_{.025}$, accept H_0 and assume the sequence is random.

14.6.3 The number of runs is $w = 19$, so $z = 1.31$.
Since $-z_{.025} = -1.96 < 1.31 < 1.96 = z_{.025}$, accept H_0 and assume the sequence is random.

14.6.5 For these data, $w = 25$, and $z = -0.51$.
Since $-z_{0.025} = -1.96 < -0.51 < 1.96 = z_{.025}$, accept H_0 at the 0.05 level of significance and assume the sequence is random.